Brock/Springer Series in Contemporary Bioscience

Protecting Biotechnology Inventions

Brock/Springer Series in Contemporary Bioscience

Series Editor: *Thomas D. Brock*
University of Wisconsin–Madison

Tom Fenchel
ECOLOGY OF PROTOZOA: The Biology of Free-living Phagotrophic Protists

Johanna Döbereiner
Fábio O. Pedrosa
NITROGEN-FIXING BACTERIA IN NONLEGUMINOUS CROP PLANTS

Tsutomu Hattori
THE VIABLE COUNT: Quantitative and Environmental Aspects

Roman Saliwanchik
PROTECTING BIOTECHNOLOGY INVENTIONS: A Guide for Scientists

Roman Saliwanchik

Protecting Biotechnology Inventions

A Guide for Scientists

With 19 Figures

Science Tech Publishers
Madison, WI

Springer-Verlag
Berlin Heidelberg New York
London Paris Tokyo

Roman Saliwanchik
Saliwanchik and Saliwanchik
Attorneys at Law
529 N.W. 60th Street
Gainesville, Florida 32607

Library of Congress Cataloging-in-Publication Data

Saliwanchik, Roman.
Protecting biotechnology inventions : a guide for scientists / Roman Saliwanchik.
p. cm. — (Brock/Springer series in contemporary bioscience)
Bibliography: p.
Includes index.
ISBN 0-910239-20-7 (pbk.)
1. Biotechnology—United States—Patents. I. Title. II. Series.
[DNLM: 1. Biotechnology. 2. Patents. QT 34 S167p]
KF3133.B56S25 1988
346.7304'86—dc19
[347.306486]
DNLM/DLC
for Library of Congress 88-18335
CIP

Science Tech Publishers, 701 Ridge Street
Madison, Wisconsin 53705 U.S.A.

Sole distribution rights outside of the USA, Canada, and Mexico granted to Springer-Verlag Berlin Heidelberg New York London Paris Tokyo

ISBN 0-910239-20-7 Science Tech Publishers Madison, WI
ISBN 3-540-50103-7 Springer-Verlag Berlin Heidelberg New York

Production and editorial supervision: Ruth Siegel

Printed in the United States of America
10 9 8 7 6 5 4 3 2 1

Contents

Foreword

On the basis of my many years of experience, first in industry and then in academia, I believe that scientists will find reading this book to be very rewarding. During my years as an industrial research scientist, I learned first-hand of the urgency of adequate protection measures at an *early stage* in the research process. I saw the rights to important inventions lost because researchers had not adequately maintained their records and files. Later, during my years as a university professor, I found it essential to teach my research students how to keep records.

However, the adequate protection of a biotechnology invention requires more than just keeping correct records. There is a whole series of procedures that must be followed. Now that biotechnology research is of such great economic importance, all researchers in this area have a vital interest in protecting the products of their research. In the present book a noted patent attorney with many years of experience in biotechnology—as both a researcher and a lawyer—presents a concise, authoritative, and easy-to-read guide to the protection of scientific inventions.

Although this book concentrates on biotechnology, the information presented is applicable to any activity in which intellectual property is created. Every researcher, whether working in industry or in an academic setting, should own and read this book, which clearly outlines the essential procedures and the common pitfalls involved in the protection of the results of scientific activity. Clearly illustrated with many flow diagrams and examples, the book expertly guides the uninitiated through the maze of patent law and trade secret protection. The author points out that if a researcher and/or an employer wishes to protect new discoveries, confidentiality must be maintained from the moment the idea is first conceived.

Although this book is written with the researcher in mind, the business person and legal technician will also find the information of value. The author, Roman Saliwanchik, is the senior partner in a Florida law firm that specializes in patent and trademark law. He was previously Senior Patent Lawyer for the Upjohn Company, and before that a microbiologist-chemist with Eli Lilly. Among his publications is *Legal Protection for Microbiological and Genetic Engineering Inventions* (Addison-Wesley, 1982). The present book offers advice that Saliwanchik has gleaned from his actual experience of over 25 years of practice in patent law. The distillation of this experience into a concise, action-oriented work will come as a boon to researchers, as well as to students, instructors, practicing lawyers, and industry executives.

Thomas D. Brock
University of Wisconsin-Madison

Preface

In the course of my work as a patent attorney for over 25 years, first at a major biomedical corporation and then in private practice, I have frequently been surprised at how little most scientists know about protecting the legal rights to their inventions. I have also witnessed the frustration of scientists who have unwittingly given up their legal rights simply because they were not aware of the proper procedures for protecting these rights. This ignorance has often led to financial loss for both scientists and their employers; it has also hindered the professional advancement of many scientists because they did not receive proper recognition for the discoveries they made.

Therefore, my primary motive for writing this book was to educate scientists about how to protect their intellectual property rights—their inventions and discoveries—and how to do so in the best way possible. In order to increase the reader's chances of successfully protecting his or her rights, I have included very specific suggestions about the proper procedures for:

- keeping laboratory records,
- maintaining confidentiality,
- making a confidential disclosure to a third party,
- disclosing data to a government agency,
- selecting a patent attorney,

and many other things that researchers should know in order to protect their work.

I have addressed this book primarily to scientists in the field of biotechnology, since this is the field in which I was trained in my first career as a scientist and also in which I practice law. As evidenced by the large number of patents and patent applications in biotechnology, this discipline is now one of the most fertile areas for new discoveries. However, I believe that this book will be very useful for scientists in other fields as well; researchers in every field should know how to protect their intellectual property rights. Ideally, all scientists should carry out their research in such a way that they do not jeopardize their rights.

In addition, from the perspective of a patent attorney, I can say unequivocally that the better informed the inventor is about the legal protection process, the more likely it is that the inventor's patent attorney can obtain protection for the invention. However, this knowledge does not obviate the need for the services of the patent lawyer. On the contrary, a well-informed inventor will be much more likely to retain patent counsel at an early stage of the patent process than will an inventor not having such knowledge.

Finally, well-informed scientists will know that the time to think about legal protection begins when an idea is first conceived and not when the invention is finally realized. They will be able to avoid numerous pitfalls and will be much more likely to receive both financial and professional rewards for their work than scientists who are unaware of the proper procedures for protecting their legal rights.

Acknowledgments

I wish to acknowledge the contributions of those who helped bring this book to fruition:

- My son and law partner David and my wife, Doris, for their editorial reviews and comments;
- Mary Jo Hoy for her typing and arranging of multiple drafts;
- A.J. Lyons, curator of the Agricultural Research Culture Collection, for information about culture depositories;
- Dr. Ruth Siegel for her editorial review and suggestions; and
- Dr. Thomas Brock for his editorial review and suggestions.

Roman Saliwanchik

Brock/Springer Series in Contemporary Bioscience

Protecting Biotechnology Inventions

1

Preserving Property Rights in Biotechnology

The scope and application of biotechnology research is extremely broad. The techniques used in this research are derived largely from the classical fields of physics, chemistry, biology, and molecular biology. The fruits of this research promise to have a profound effect on numerous disciplines including pharmaceutical, agricultural, medicinal, and environmental sciences.

Although biotechnological research has garnered a tremendous amount of attention in recent years, and the prospects for future developments are exciting, this field of research is by no means new. Biological organisms have been used for centuries for fermentation, degradation of wastes, and many other tasks. The utility of microorganisms in making wine, sauerkraut, soy sauce, and other fermentation products has long been recognized. Despite this long and illustrious heritage, it has only been within the last fifty years that biotechnology has emerged as an industrial tool of monumental significance. The origins of this emergence can be traced back to the advent of antibiotics during and immediately following World War II. After the war, antibiotics rapidly became an enormously important, and profitable, health care product. The so-called antibiotic era that followed lasted for at least 25 to 35 years. Although the production of antibiotics by fermentation is still the backbone of a number of large pharmaceutical companies and generic drug supply operations, there is little dispute that the antibiotic era is waning.

One development from the early days of antibiotic research accelerated the growth of this industry beyond the level attainable by use of fermentation technology alone. This was the discovery and development of biologically inactive moieties which could be prepared by fermentation means and then chemically converted to biologically active antibiotic compounds having unique and desirable medicinal properties. Scientists who discovered the presence of 6-aminopenicillanic acid (6-APA) in penicillin G fermentation beers laid open a whole new assortment of second-generation antibiotic compounds which quickly achieved success in antibiotic therapy.

Antibiotic inventions provided the impetus for many research and industrial concerns to grow and prosper. The protection of the rights to inventions was of vital importance to these concerns. Many of the inventions involved the use of microorganisms which were akin to highly efficient factories in making new and useful products. Also, new products in the antibiotic and steroid sciences were being discovered and brought to the market place at a rapid pace. There were a multitude of inventions which required new legal procedures for developing and maintaining the legal rights of the inventors. One result of this tumultuous period of research and pursuit of patent protection was the creation of a large body of law dealing with many complex situations. Some of the many patents which have been issued for biotechnology inventions are listed in Appendix C.

Much of the research and development work in the era of antibiotics and steroids was carried out in industrial laboratories, or in well-funded academic laboratories. Legal rights were pursued within large industrial settings where legal services were readily available. Corporate lawyers were accessible from the beginning of projects to advise scientists, to set up protective procedures, and to procure legal rights to the inventions. The net result was that the majority of legal rights established in those "early" areas of concerted biotechnology research and development, inured to industry, and also, in some cases, to well-funded academic laboratories.

Both the scientific community and the legal system have benefited from this activity. Many rules and guidelines were formulated for future guidance for obtaining legal rights in new inventions. Without this expenditure of time and resources by the industrial laboratories to establish legal rights, the scientific/legal community today would be forced to operate in a state of uncertainty. Despite the legal foundations which have been laid during the past 40 years, there are still areas of uncertainty as to whether certain biotechnology inventions can or cannot be protected by certain legal procedures. Today's "newer" biotechnology has many complexities which do not readily fit into the rules and guidelines of the older biotechnology. Thus, there remain many areas of uncertainty for the scientist and patent attorney to consider before deciding upon the best route for obtaining sound legal protection for biotechnology inventions. As the following chapters will show, unique, creative solutions are often required in order to realize a sound legal position.

Despite the new aspects of biotechnological research, many of the research fact situations attendant this newer biotechnology are analogous to the research settings of prior biotechnology. Thus, procedures designed to ensure the legal protection of research results in prior biotechnology work are often applicable to the newer research.

However, rapid changes in biotechnological procedures have produced special problems related to maximizing the benefits that individual scientists and their employers can be expected to realize at various junctures in their work. It is well known in the scientific community that those who produce significant advances are not always the ones who receive the rewards. Though the overall system is working to recognize the achievers in biotechnology research, there remain pockets of unfairness; instances where the true achiever is never recognized as such, or where a later entry gains the reward.

It is unrealistic to believe that any system of recognition and rewards for scientific achievement would function at all times without defect. History records many disputes among those claiming to be the first to contribute to the sum of human knowledge. Yet there are procedures which can be used by scientists which can give the scientist a better chance in realizing recognition and rewards. Chapter 3 provides specific procedures which will help scientists reach these goals.

A large number of scientists, perhaps even a majority in some research environments, do *not* have legal information which is necessary in order to be recognized and rewarded for research work. Their knowledge may be adequate for the scientific work they do, but they are utterly lacking in knowledge of how to protect their work product from being pirated by others. One major goal of this book is to provide the scientist with the knowledge needed to work with a lawyer in an effort to preserve invention rights and the attendant recognition and rewards. This knowledge is especially important at the earliest stage of the scientist's work, at a time when ideas are first conceived to overcome or rectify a particular problem. It is at these early stages that many inventors unknowingly lose their rights. The result can be many subsequent years of unrewarded work to bring the idea to reality.

Therefore, in order to avoid unfortunate forfeiture of rights, while maximizing the chances of obtaining legal protection and professional recognition, it is crucial that scientists follow proper procedures at all stages of their research activities. From the initial formulation of an inventive idea until legal rights are firmly established, researchers must be cognizant of legal pitfalls and the procedures which can be followed to avoid these pitfalls. The purpose of this book is to describe in detail what these procedures are. Though, in a general sense, much of what follows can be used to guide scientists in other fields as well as those in biotechnology, procedures relating to problem areas of biotechnology will be emphasized.

2

Two Forms of Legal Protection

Patent protection and trade secrets are the two most important means of protecting property rights in biotechnology inventions. The purpose of this chapter is to introduce these alternative forms of legal protection. A brief introduction into the concept of "invention" will help to lay the groundwork for the ensuing discussion of patents and trade secrets.

Webster defines invention as "something invented." To invent is "to create or produce for the first time." Thus, in a general sense, an inventor is someone who has created or produced something for the first time. Although many people automatically associate the word "invention" with patents, the fact that something has been done for the first time, i.e., an invention has been made, does not mean that the invention is *patentable*. We all know that patents are granted for inventions; however, the truth of the matter is that many inventions are *not patentable*. A particular invention may not be patentable for any number of reasons depending on the particular laws of the country where a patent is being sought. The point here is that an "invention" is the product of a creative act, whereas patents are legal documents which attest to the fact that the invention covered in the patent meets the test for patentability in the country issuing the patent.

The United States has a Patent Act which delineates the requirements that an invention must meet before a United States patent will be granted. In addition, all the major countries of the world have their own patent laws to govern the granting of patents in their countries. This means that a particular invention may be patentable under the laws of one country but not under the laws of another country. Generally speaking, the United States patent law is the most comprehensive one in the world. Many countries have followed the United States law to give greater patent coverage for inventions. Appendix D contains excerpts of relevant sections of the U.S. patent law code.

2.0 Patents—United States and Worldwide

Patents are given by governments to inventors for inventions which meet the statutory requirements of the particular government. An invention which may meet the requirements of one country, and, thus, be considered to be "patentable," may not meet the statutory requirements of another country. This means that when decisions are made pertaining to the patenting of an invention, the decisions must take into account the requirements of the particular country from which a patent is desired. An inventor residing in one country can apply for a patent in other countries. Once a patent is granted by a country, the legal rights attendant this act are enforceable only within the confines of the granting country and its territorial possessions. Therefore, an inventor needs to obtain a patent in each country where he or she wishes to have protection.

In the United States the basic tenets of patent law are set out in Title 35 of the United States Code. This statutory enumeration of the requirements for obtaining a patent is augmented by United States Patent and Trademark Office (PTO) rules and regulations. The Patent statutes and the PTO rules and regulations are augmented further by legal decisions handed down by a variety of federal courts. This entire complex of statutes, rules, regulations, and decisions make up what is referred to as the "patent law" of the United States. The basic statutory requirements for an invention to be patentable under the United States Patent Act is that the invention be (1) *new,* (2) *useful,* and (3) *unobvious*. Although this may sound simple, the determinations under these requirements can be very complex.

If an applicant is able to demonstrate to the PTO that his or her invention meets these and other statutory requirements, then a United States patent will be granted to the inventor. As a general rule, the granting of a United States patent gives the inventor the exclusive right to make, sell, or use the patented invention for a period of seventeen years from the date of issuance of the patent. Although there are some qualifications to this general rule, such as the so-called "experimental use" exception which will be discussed in Chapter 7, the granting of a patent clearly conveys important and valuable rights to the patentee.

Although the benefits bestowed on patentees are substantial and are immediately apparent, it is important to recognize the many benefits which accrue to all of society as a result of the patent system. The primary *raison d'etre* for the United States patent system is the promotion of science which, in turn, results in a better economy and standard of living for all citizens. The Founding Fathers of the United States recognized that scientific research and development could best be promoted by rewarding those people who are willing to invest time, effort, and money to create new and useful inventions. One crucial element of the patent system is that the inventor must give a full disclosure of how to make and use his or her invention; no patent is granted unless such a complete disclosure of the invention is made. The patent document, which becomes public as of the date of issuance of a patent, discloses enough information about the patented invention to enable a person skilled in the art to which the invention pertains to repeat or practice the invention on an experimental basis. This opportunity to learn about another's creative work has been shown time and again to foster additional creative work by other researchers. The net result is a "snowballing" effect wherein new inventions continue to build upon earlier ones. The limited monopoly granted to patentees is simply the small price that society pays in return for the inventor's full disclosure of his or her new idea.

Without an efficient patent system, potential inventors would have much less incentive to spend the time and money necessary to create new and useful ideas. Furthermore, inventors would have no incentive to disclose their new idea to the public. Instead, they might use the invention for their own personal needs, or, where possible, they might try to market their invention while keeping the details of their invention a secret in order to maintain their market position. Clearly, under this scenario, scientific research would never have reached the level which has been attained under our patent system.

Virtually all of the industrialized countries of the world, and certainly all of the major industrialized countries, have similar patent law structures. There are, as would be expected, differences in form and procedure based on the particular legal history of a country. Differences in the substantive aspects of the patent laws exist, but, generally, there is a high degree of homology in the substance of the patent laws of the major countries throughout the world. This homology is particularly marked in the patent laws of the English-speaking countries. And, perhaps, this is to be expected because laws other than patent law are also similar among such countries.

Yet, because differences do exist in both form and substance in the patent laws of closely related countries, any assessment of the patentability of an invention must be made in the context of the laws of the particular country in question. Such an assessment can only be made intelligently by someone with a working knowledge of the patent laws of the jurisdiction involved. This all leads to the obvious conclusion that an expert in the law of a particular country must be involved in making the patenting assessment.

Because of the many possible barriers to obtaining a patent, the patenting process should be carried out by a skilled patent practitioner, generally referred to as a patent attorney. Every licensed patent attorney has a scientific background and has passed an examination testing the attorney's knowledge of patent laws. The complexities of the various scientific disciplines necessitates a certain amount of specialization even within patent law. For example, if an invention concerns biotechnology, then the services of a patent attorney skilled in biotechnology and patent law should be utilized. Though a patent attorney with a mechanical or electrical engineering background might also function suitably in patenting a biotechnology invention, it is more likely that a patent attorney trained in biotechnology could readily understand the scientific terminology and be familiar with the subtle nuances of drafting applications involving that technology.

2.1 Trade Secrets

Trade secret protection is a creation of common law and, in some instances, state law. Unlike patents, there are no federal statutes granting trade secret protection. Consequently, trade secret law may vary in some respects from state to state. Despite this potential for slight variations, the fundamentals of this form of protection are well established and followed by each jurisdiction. A trade secret can be a viable form of legal protection for both non-patentable and patentable inventions. A non-patentable invention is one which does not meet the requirements of the patent law for a particular country. In order to facilitate this discussion, I will concentrate primarily on United States law and make reference to foreign laws only if they differ greatly from United States law.

Trade secrets are often defined as "any formula, pattern, device, or compilation of information which is used in one's business, and which gives him or her an opportunity to obtain an advantage over competitors who do not know of it." The subject matter of a trade secret must be secret. Examples of things which could be held as trade secrets are: a formula for a chemical compound; a process for making, treating, or preserving a product; and a microorganism. As stated previously, a trade secret can be an invention that is either non-patentable or patentable.

Specific examples of trade secrets are as follows:

1. A process to brew beer using a specific yeast strain.
2. A chemical for use in a fermentation of an antibiotic that gives higher yields of the desired antibiotic.
3. A process condition such as temperature or aeration rate, which leads to a higher yield of a desired product in a fermentation.

The above trade secrets may or may not be patentable. For example, if the secret is not novel or if it would be obvious in view of what is already known by a person skilled in the art, then a patent could not be obtained.

From what has been said so far about trade secrets and patents, it should be evident that the two forms of protection

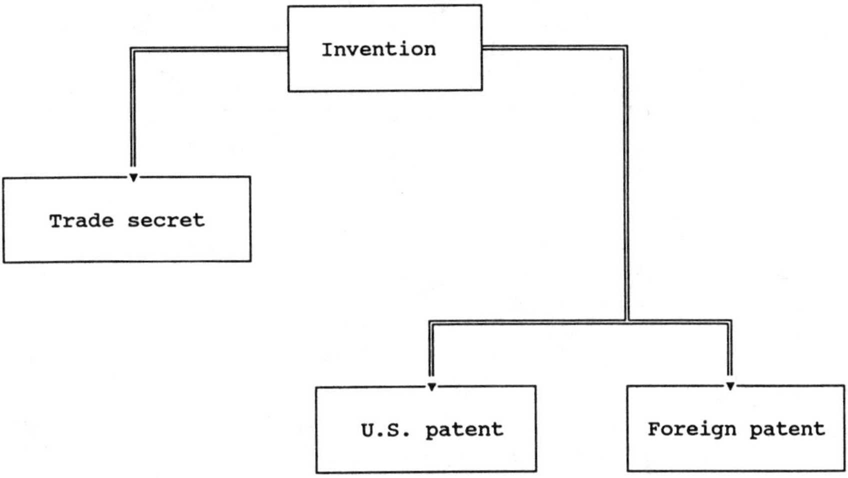

Figure 2.1 The primary options for protecting inventions.

are mutually exclusive: patents result in a full public disclosure of the subject matter, while trade secrets require that the invention be kept in confidence. If an invention is determined to be non-patentable, then trade secret protection is probably the only choice for legal protection of the invention. However, if an invention is considered to be patentable, then, a choice will have to be made between trade secret and patent protection. Figure 2.1 illustrates these options. The final decision will be dependent upon a detailed analysis of the invention facts with regard to the relative advantages and disadvantages of trade secret and patent protection. An account of the characteristics of trade secrets and patents that are used to make such a decision will be given in Chapters 8 and 9. However, the inventor should engage a competent patent attorney to analyze the facts and to make the decision.

3

Protecting Intellectual Property Rights

The law recognizes a person's right to dominion, or control, over a variety of subject matter. Virtually everyone is familiar with property rights relating to real estate and personal property. Conceptually it is not difficult to understand a person's right to exclude others from his or her land, or to sell livestock or a painting. However, there are other forms of property which may be just as valuable but are less tangible. One example of an intangible property right is ownership in a corporation, as evidenced by a stock certificate.

Among the more esoteric forms of property are copyrights, patents, and trademarks. These forms of property, along with several related property forms, are collectively called intellectual property. The name "intellectual property" is derived from the notion that these rights arise primarily as the result of creative cerebral activity on the part of an individual or group of individuals. Because our society and system of laws has never required people to disclose their ideas, each person is free to exercise control over at least the initial disclosure of his or her idea. The United States legal system, as well as the legal systems of virtually every developed country in the world, recognizes certain property rights that an individual has with respect to his or her creative ideas. However, in order to reap the potential benefits from the sale or use of this intellectual property, there are certain steps that the holder of the property must take in order to protect these property rights.

Although the typical member of society may seldom, if ever, have to be concerned about the protection of intellectual property rights, scientists should always be cognizant of the legal ramifications of their research. Because of the nature of a scientific researcher's profession, intellectual property rights could account for a large portion of his or her personal assets. Therefore, it behooves the researcher to be well versed in the procedures which must be followed in order to protect intellectual property rights.

There are four principal means by which intellectual property may be legally protected: trademarks, copyrights, trade secrets, and patents. These forms of legal protection are primarily distinguishable on the basis of the type of subject matter that each protects. For example, trademarks are used to protect property rights in unique names or slogans which serve to identify a particular product or service. Copyrights are used to protect literary and artistic works such as books, paintings, sculptures, and sound recordings. Trade secrets and patents, on the other hand, are used to protect property rights in new inventions. Because researchers in general, and biotechnology researchers in particular, are primarily interested in patent and trade secret protection, this book focuses on these two types of legal protection. The protection, in a legal sense, of intellectual property rights should be practiced continuously from the initial stages of any research project to its end. The initial the research that the researcher conceives. When a researcher first originates an idea for solving some existing problem, even this idea itself should be sheltered from public view until the individual has taken certain steps to preserve the fact that the idea belongs to the researcher. In addition, if in the course of working on an existing project, a researcher observes an event or result that leads to an idea for solving another problem, this new idea should also be handled in a manner which will preserve its legal identity as the researcher's intellectual property.

One of the primary reasons for taking these early steps to protect intellectual property rights is that as a general rule, under the United States patent system, when the Patent Office determines that an invention is patentable, the patent is awarded to the first person who came up with the idea. This system is in contrast to the system used by many foreign countries where patents are awarded to the first person to apply for a patent. In countries following the first-to-invent system such as the United States, the date of conception of an idea is of critical importance.

The United States Patent and Trademark Office uses an *interference proceeding* in order to resolve disputes relating to first inventorship. Although interference proceedings involve complicated legal procedures which are beyond the scope of this book, the objective of the proceeding is simply to determine the first inventor of a new idea. As one would expect, research records often play a central role in these determinations. For this reason, the protection of intellectual property rights should start in the initial stages of any research project.

Many scientists find that the tendency to immediately discuss new observations, results, or ideas with other researchers is irresistible. Though these immediate discussions do not usually affect the legal determination of inventorship, such discussions have resulted in disputes over the ownership of the initial idea in the case of many significant scientific contributions.

In the next section I will discuss, in detail, procedures for preserving the legal inventorship of inventions. Following these steps will ensure that researchers will subsequently be able to establish their ownership of inventions. These steps are not burdensome; they consist of simply documenting events in a manner that is legally acceptable.

3.0 Documentation of Research

Before discussing the research work itself, let us take one step back to the initial idea stage of a project. When a researcher first has an idea for solving a problem, he or she should record the idea in a bound laboratory notebook as soon as possible. This should certainly be done before discussing the idea with anyone other than the researcher's immediate family. If the idea occurs to the individual at home, then a member of the household can sign the notebook page as a witness of the event, after the person having the idea signs and dates the notebook page. If the idea arises at the workplace, then a laboratory assistant can witness the idea as above. But, whether at home or in the workplace, even after the idea has been duly recorded and witnessed, further disclosure of the idea should only be made to persons who need to know about it. For example, in the workplace, the researcher would normally report the idea to a superior. By immediately recording the idea, the researcher is preserving the identity of the source of the idea and, also, setting in motion the procedures beneficial to the employer for preserving legal rights to the idea and any results emanating therefrom.

This procedure of documentation of ideas prior to disclosure to others is not simply an idle exercise; it can have significant practical consequences. Because of the tendency to discuss ideas with co-workers and even researchers at other institutions, questions often arise as to which researcher actually had the idea first. It is not unusual for the recipient of such communications to subsequently become identified as one of the originators of the idea. That is, the recipient may take steps to be regarded as a participant in the origination or development of the idea. This is particularly true if the idea appears to be of potential commercial worth. Thus, immediate recordation of ideas helps to clarify honest confusion as to inventorship, and it may also discourage the recipient of information from attempting to be identified as an originator of the idea. At the very least, recordation of ideas will help a patent attorney prove the true identity of the inventor in subsequent legal proceedings.

If ideas originate as part of a group discussion at a research meeting, then the group may be considered the originator of the idea. The group as a whole may be considered the inventor if, for example, the cross-fertilization of ideas by members of the group cannot be separated from the ultimate expression of the idea. In such cases, the group should record the idea immediately, as discussed above.

With regard to the recordation of data and observations, it is important for scientists to recognize that it is just as important to document work that gives a *negative* result as it is to document work that gives a positive result. Generally, results are either positive or negative based on the particular design of the work plan and the desired result. In the case of negative results, it is possible that they also may lead to significant inventions. For example, in the initial screening of soil samples for antibiotic-producing microorganisms, one often finds microbes which produce substances that show only marginal activity against test microbes. The marginal activity may be attributable to the absence of any compound with strong antibiotic properties. However, it is also possible that a strong antibiotic is present but its activity is attenuated due to the presence of an interfering compound or due to its own low concentration. In either case, further work might yield the isolation of a new strong antibiotic compound. By recording the marginal activity, the researcher leaves open the possibility of going back and doing further work with the microbe. It is true that the realities of research economics may not allow further work to be done immediately on anything but cultures giving the best, or most positive, results. However, this is not a valid reason for not documenting negative or marginal results.

The primary purpose of recording research results is to prove that particular specific research was done at a certain time by a particular person. In order for the recordation to effectively achieve this purpose, the documentation must be legible, done in pen, and entered into a *bound* laboratory notebook. Any deviation from these three minimum requirements may cast doubt on the accuracy or authenticity of the record. For example, the bound laboratory notebook ensures that no one has removed or inserted pages in the notebook. Proper documentation of research work helps to ensure that a court in an interference proceeding will be cognizant of *all* the relevant work which was done by a researcher.

Thus, if researchers consistently practice the following suggestions, it will be easier for them to establish their rights to a particular invention.

Specific recommendations for the best way to keep a record of research work are as follows:

1. Use a bound laboratory notebook with numbered pages to record all experiments.
2. Write the date that the work started in the upper right corner of the page.
3. Make all notebook entries in ink of a single color to facilitate proving the content of the entry.
4. Fully describe the planned details of the experiment indicating the purpose of the experiment and the expected result.

5. If some of the experimental work is done by another person, for example an assayist, the data obtained from the assayist should be recorded in the bound laboratory notebook as soon as they are received by the researcher who submitted the material that was assayed.
6. Write the date that the assay results were received on the same page where the results are recorded, and identify the assay date by reference to a prior notebook page that contains a description of the material upon which an assay was requested.
7. Make entries directly into the notebook as the experiment is carried out and immediately when results are obtained.
8. If an incorrect entry is made, draw only a single line through the incorrect entry; there should be no attempt to obliterate or erase the incorrect entry.
9. Enter all conclusions and observations about the experiment into the notebook; even negative experimental results are important and, therefore, should be included in writing in the notebook.
10. Immediately after each page of the notebook is completed, it should be initialed and dated by the person actually doing the experiment; this person may very well be a laboratory assistant.
11. Each page of the notebook should also be initialed and dated by the person who supervised the person actually conducting the experiment.
12. The recording of the conclusion of an experiment should be followed immediately by the signature of the experimenter; and a witness, i.e., a person who observed and understood the experiment or a person to whom the experimenter disclosed the experimen-

tal details and who understood the experiment, should sign thereafter.

13. If a notebook page is not completely filled out, a diagonal line should be drawn through the blank portion.

Data entered into a notebook should be accurate and legibly written. In-house jargon that is not commonly known in the particular art of the experimenter should be avoided or thoroughly explained. For example, reference to an in-house assay procedure by the term "Joe's assay" is without probative value, i.e., it does not provide proof or evidence, standing alone. Therefore, the details of the assay should be set forth in the experimenter's notebook. If the same assay is used frequently, then the first entry of the assay details should be referred to when the assay is repeated. Also, the reference to materials by their trademarked name should be accompanied either by reference to the source of the materials or by a description of the composition of the materials. In summation, the notebook record should be complete enough so that any person who is reasonably skilled in the particular art of the experiment will be able to understand the procedure used.

The details of experiments can also be documented by in-house memos as well as reports to supervisors and associates. Therefore, such reports should also be of sufficient detail to enable other researchers to understand the procedure discussed in the report.

3.1 Maintaining Confidentiality

An inadvertent loss of confidentiality of research work can jeopardize efforts to obtain patent or trade secret protection, and it can allow competitors to get valuable information.

Therefore, researchers must take steps to protect the confidentiality of their work until the time comes when legal protection is in place and a decision is made to publish or otherwise disseminate the work. From the very beginning of a research project, scientists should take precautions to prevent unwanted disclosures of their work.

A key part of this confidentiality is to limit access to research notebooks. A researcher's notebook should not be left unattended in an area that is accessible to persons outside the researcher's immediate work group. This means, for example, that a research notebook should not be left unattended on a desk or laboratory workbench if the researcher and/or assistants are not present and the area can be legitimately entered by other people. This situation can occur, for example, over a lunch hour or seminar break. To prevent this, the researcher should either lock the office or laboratory door, or lock the notebook in a desk or cabinet. To some people this may seem to be counter to an open research environment where the exchange of ideas and results is considered desirable. However, the research environment itself is not adversely affected when researchers preserve the confidentiality of their work from outsiders. It certainly does not advance the research cause to have a person outside the research team look at, or even take, research notebooks without the knowledge and permission of a member of the research team. If an outside person, even an employee of the same company, wishes to review or discuss research work done by another person or group, then such outside person should make the proper request of persons directly involved with the work.

Whether or not a supervisor in a company should have access to a researcher's notebooks at any time depends upon the operating standards of that workplace. If a particular supervisor has worked closely with a research group and is, in effect, considered a member of the group by virtue of a continuing dialogue with it, then there is no loss of confidentiality if the supervisor peruses a notebook. However, it may be counterproductive to allow any supervisory person, aside from an immediate supervisor, access to all research notebooks.

Generally, the primary goal of any research program is to produce results that will benefit the researcher, the employer and eventually, the public. Protection of laboratory notebooks can be of great importance in the pursuit of this goal. Accurate and complete research records facilitate efficient protection of legal rights. For example, if all notebooks are properly protected, it is unlikely that a missing notebook will block attempts to prove an invention date. Also, when the work done by the researcher is protected by keeping it as a trade secret rather than by filing for a patent, as discussed in Chapter 2, maintaining the confidentiality of the notebooks is critical to the subsequent enforcement of the trade secret. This is because courts will not enforce a trade secret if the owner of the trade secret cannot show that proper steps were taken to preserve the trade secret. Clearly, leaving notebooks unattended is not a proper step for maintaining a trade secret.

3.2 Reporting To Associates and Colleagues

Almost all scientists want to communicate the results of their work to their peers as soon as possible for recognition, approval, and constructive criticism. But, if they do so prematurely, they can lose all legal rights to the research work. Premature communication occurs when the communication precedes the establishment of a legal position. However, if the communication is directed to a limited group of associates within a well-defined area in a company or research setting, then such communication does not normally constitute an impediment to obtaining legal rights. This type of disclosure is least likely to cause a problem when the group receiving the communication is under some type of obligation to the company or research group to maintain the confidentiality of the information communicated to them.

Communication of research work to associates can be in the form of discussions, seminars, or written reports. Obviously, much can be gained from discussions with associates, and, therefore, such communications are desirable. However, it is advisable to limit the number of persons receiving confidential communications. Only persons who can contribute useful ideas or have a particular need to be aware of the subject matter should be invited to seminars. Written reports should be sent to only these people as well. There should be no recordings or unnecessary copies made of communications. Also, care must be taken not to leave copies of a report in a place accessible to people outside the research group. Leaving a copy on a seminar table or in a library can lead not only to a loss of legal rights to any invention described in the report, but also to a premature disclosure to the public of confidential research work. Furthermore, the public, in many instances, can include a competitor.

In a relatively small research setting, control of in-house communications in order to maintain confidentiality can be very efficient. However, in a larger research setting, such as a large corporation, communicating to the in-house group is often more akin to communicating work to an outside group. The distinguishing feature is that the chances of a loss of confidentiality increase with the number of people who are privy to the information. A real-life example illustrates how confidentiality can never be taken for granted.

In the course of an accelerated research program to develop a new antibiotic, a famous drug manufacturer decided to concentrate heavily in a very confidential area of research. The company took various steps internally in order to maintain the secrecy of the project and believed that confidentiality was being maintained. It came as a shock, therefore, when a

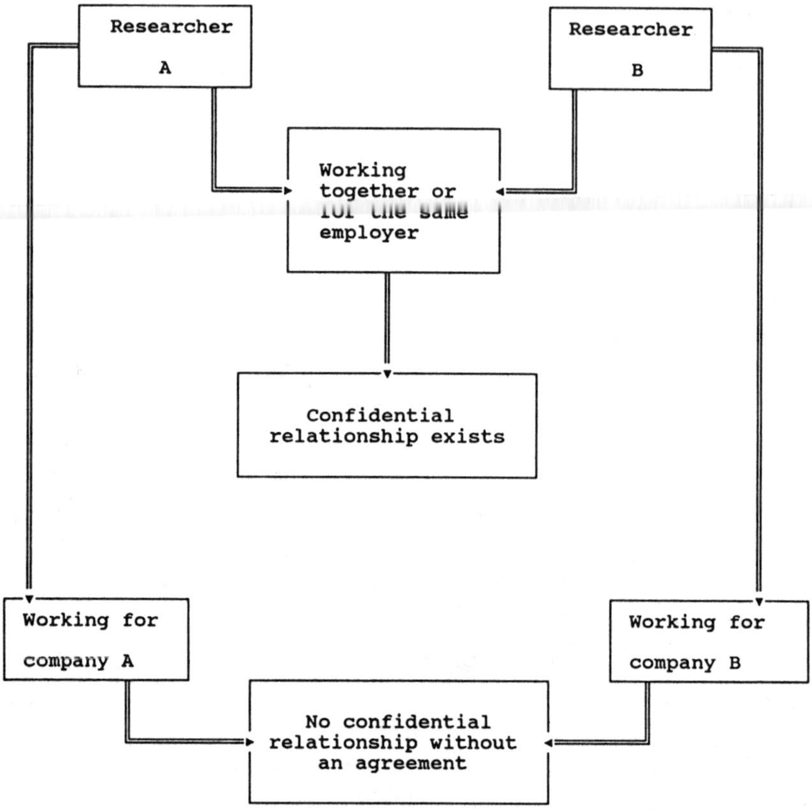

Figure 3.1 The confidential relationship between two researchers who work for the same employer is compared to the relationship between two researchers who work for different employers.

senior research executive, on a routine personal trip to his drug store, was asked by the pharmacist, "Say, Doc, when are you expecting to go for an NDA [New Drug Application] on your new antibiotic X? I've heard that the results are great against A, B, and C bugs." Of course the executive was very upset to find out that the research was not secret after all. Clearly, large corporations can be difficult environments in which to control the dissemination of research information.

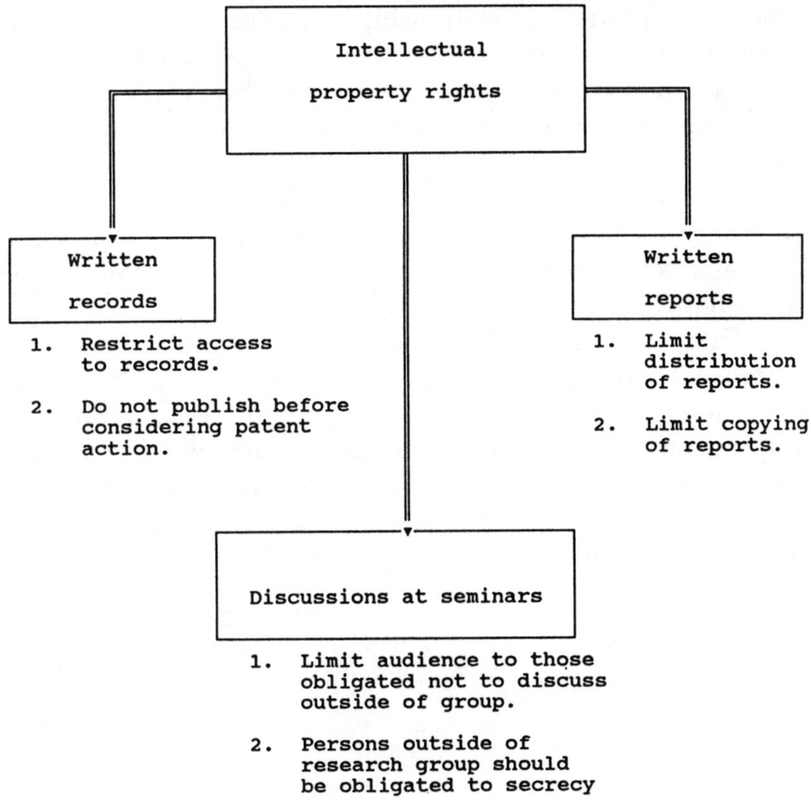

Figure 3.2 Protecting intellectual property rights: A summary of the steps researchers should take to maintain the confidentiality of their work.

So far, the discussion has focused primarily on "in-house" communications. Disclosures to people outside of a research institute pose additional confidentiality problems. Once a researcher decides to seek peer review outside the confines of the researcher's work setting, that is, from outside scientists, the risk of losing confidentiality increases and additional precautions should be taken. The most important of these precautions is to limit dissemination to those individuals with whom a confidential relationship has been established. The confidential relationship must be established *before* any communication is made. Figure 3.1 compares the relationship between two researchers who work for the same company versus two who work for different companies. No communications should be made in the latter case unless a confidential disclosure agreement is executed between the two parties. Confidential disclosure agreements are discussed in detail in the next chapter.

Thus, limiting and controlling the communication of research is absolutely necessary for protecting legal rights for inventions. There should be some communication in most instances, but this should be limited as much as possible, and all communications should be done in a confidential setting. Figure 3.2 summarizes the steps researchers should take to maintain the confidentiality of their work.

4

Establishing a Confidential Relationship

Before a researcher discloses confidential information to any person outside his or her research group, steps should be taken to insure that the recipient of the disclosure will keep the information in confidence.

If the person receiving the disclosure, that is, the disclosee, is an employee of the same company, then his or her employment contract probably requires keeping such disclosures confidential. If this is the case, the only steps that need to be taken are steps designed to impress upon the disclosee that the information is, in fact, confidential. This can be accomplished by explicitly stating orally or in writing that the information must be kept in confidence. Also, all documents associated with the disclosure should be clearly marked as "Confidential." These precautions will remind the disclosee of his or her duty to keep the information in confidence.

Frequently, researchers or their employers must transmit confidential information to people outside the company. The purpose of such disclosures may be, for example, to obtain funding, to have field tests run, or to have chemical structures determined. Naturally, the discloser wishes to transmit information without fear of the disclosee using the information to his or her own advantage or passing the information to competitors. In order to prevent unauthorized use or transmission of confidential information, the discloser of the information should establish a confidential relationship with the disclosee. The primary means for establishing this confidential relationship involves the use of a "Confidential Disclosure Agreement." This type of agreement is relied upon heavily throughout the academic and the industrial world to transmit ideas and information from one entity to another.

A confidential disclosure agreement is a written agreement between the person disclosing the information and the person receiving the information. Once both parties sign the agreement, it is a legally binding contract. Only after this agreement has been signed should any disclosure be made. This mechanism of using a confidential disclosure agreement is convenient and readily adaptable to the facts of a particular situation. If the agreement is drafted and executed properly, it can provide the legal framework for protecting research work prior to obtaining formal legal protection, such as a patent. If there is a breach of the agreement by the party receiving the confidential information, then the usual legal remedies for breach of contract are available. For example, a suit can be filed in a state court seeking damages and/or an injunction to prevent additional harm.

4.0 Confidential Disclosure Agreements

Although confidential disclosure agreements are very adaptable to the subject matter involved, and they exist in a variety of formats, there are certain elements which are common to virtually all such agreements. Figure 4.1 is an example of a confidential disclosure agreement between a researcher and a corporate entity.

One of the elements common to all confidential disclosure agreements is the identification of the parties to the agreement. Also the agreement should include a description of the information which will be transmitted to the receiving party. Of course, the *details* of the information will only be conveyed *after* the agreement is signed by both the discloser and the disclosee. Therefore, the brief description contained in the confidential disclosure agreement should *not* be so detailed as to disclose confidential aspects of the subject matter. The reason for this is that the agreement itself will be received by the disclosee *before* a confidential relationship has been established. It is important to note that the desired confidential relationship is established only after the agreement is signed by the discloser and the disclosee. The actual disclosure of the confidential subject matter must wait until confidentiality is established. Therefore, the brief description contained in the confidential disclosure agreement should be detailed enough to identify the subject matter, but not so detailed that confidential information is disclosed. For example, if the confidential information concerns a new process for making aspirin, then the agreement could state that the confidential matter being disclosed concerns "a microbiological process using a new microbe to make aspirin." Generally, this will be enough of a description to identify the subject matter, but certainly not enough to divulge the content of the process which is the essence of the secret.

If a particular disclosure is in a crowded area of technology, for example, a new fermentation process used to make penicillin, then a more detailed identification of the subject matter of the agreement would be required to distinguish the secret from the previously known processes. This kind of identification could take the form of "a fermentation process for

Confidential Disclosure Agreement

SHOE DRUGS, INC., (hereinafter referred to as DISCLOSEE), a corporation of the State of Delaware, with offices in Gainesville, Florida, is willing to accept a disclosure from:

DR. IRA ERICKSON

(hereinafter referred to as DISCLOSER) relating to:

A Microbiological Process for Preparing Aspirin

SHOE will hold the novel aspects of DISCLOSER'S disclosure in confidence for five (5) years from the date of receipt. The novel aspects of DISCLOSER'S disclosure are those which:

1. Are not already known to SHOE from their own research or from a third party who has a right to disclose the information to SHOE; or
2. Are not or do not become subject of a patent issued to a third party; or
3. Are not or do not become open to the public by any means including disclosure in patents or publications, or public use or sales.

SHOE hereby agrees not to make or sell the disclosed idea without the express written permission of DISCLOSER. Further, SHOE hereby agrees not to pass the idea to any third party without the express written permission of DISCLOSER.

SHOE agrees to review the disclosure and notify DISCLOSER within thirty (30) days if SHOE wishes to obtain an interest in the subject matter of the disclosure.

SHOE and DISCLOSER have no further obligations with respect to the disclosure.

______________________	______________________
Representative of SHOE	DISCLOSER
______________________	__________
Representative's name	Date
______________________	
Representative's title	

Figure 4.1 An example of a confidential disclosure agreement.

preparing penicillin using a new class of chemical compounds as precursors." The point again is that sufficient identification of the subject matter should be given without divulging the secret *per se*.

The confidential disclosure agreement should also specify how long the information will be held in confidence. Generally, the time period will run from about three years to about ten years. This time period is somewhat dependent upon the contemplated life of the subject matter of the agreement. If the subject matter is in a scientific area which is changing rapidly, then, realistically, a time period of about three years would be sufficient.

Most confidential disclosure agreements contain a clause which relieves the disclosee of its duty of confidentiality *if and when* the confidential subject matter ever becomes generally available to the public. For example, if the confidential disclosure agreement specifies a ten year period of confidentiality, but a patent issues on the subject matter after six years, then the agreement would not be enforceable against the disclosee for the last four years. This is so because the agreement covers only the discloser and disclosee and does not cover the actions of anyone not in a confidential relationship with the disclosee.

Another clause included in many confidential disclosure agreements specifically prohibits the disclosee from passing the confidential information to others. It is advisable to state in the agreement that the information will not be disclosed to others by the disclosee unless others need to know the information in order for the disclosee to make a proper evaluation of interest in the subject matter. This situation can arise where a disclosure is made to Dr. X, and Dr. X needs to have assistant Y perform certain tests in order for Dr. X to ascertain interest in the subject matter. Disclosure of the information to assistant Y would not be a breach of the confidentiality agreement. In this case it is best if Dr. X and assistant Y have a confidential work relationship, which should be documented by a written agreement. Such an agreement is desirable in any research setting so that research assistants fully understand their responsibilities with regard to maintaining the confidentiality of all work performed in that setting.

Even if a confidential disclosure agreement does not specifically state that the disclosee cannot disclose the subject matter to others, it is generally recognized in the law that such a prohibition is a part of the agreement. However, for ease of enforcement, an express statement is recommended.

Even though a discloser believes that the ideas or information that he or she is disclosing are new, there remains the possibility that the disclosee already knows about the subject matter, or that such knowledge is already available to the disclosee in the disclosee's work environment. In such cases, the disclosee cannot be obligated to any confidentiality with respect to such specific subject matter. Proof of such prior knowledge by the disclosee requires that the disclosee produce a properly executed written document that is dated prior to the date when the disclosure of the information occurred. Such a suitable document could be a bound laboratory notebook describing the subject matter of the disclosure. The notebook entry should be properly dated, signed, and witnessed, as described in Section 3.0.

In this regard, it is recommended that *before* a disclosee receives a disclosure from an outside source, the disclosee should update all pertinent research records in the disclosee's work setting. This updating should take the form of entering all pertinent research work into bound laboratory notebooks in the proper fashion. "Pertinent" research work is any work related to the general area of the contemplated disclosure. For example, if the disclosure will concern a fermentation procedure for making penicillin, then the disclosee should make sure that all research (and development) work concerning penicillin fermentation processes is brought up to date in bound laboratory notebooks. If this is not done, then it could be very difficult for a disclosee to later prove having prior knowledge of a disclosure.

This factor of prior knowledge is a difficult issue for both the discloser and disclosee if the disclosee is a large corporate entity having multiple research laboratories where research projects overlap. Because of this factor, large corporations are very careful about receiving disclosures from outside researchers so as not to "contaminate" their own research and development work. Agreements with disclosees that have large multiple research units often have to be tailored specifically and precisely in order to recognize the relationship among such units.

As a follow-up to a meeting of the discloser and the disclosee during which confidential information is disclosed, some parties may exchange written communications within a few days to acknowledge the identity and extent of the disclosure. Unfortunately, such an after-the-fact system of communication frequently, if not always, lessens the cooperative atmosphere initially intended between the parties. This may be the result of a disagreement as to what the discloser actually disclosed to the disclosee during their verbal communication. Naturally, the disclosee will take the position that the discloser disclosed less information than the discloser says was disclosed. This is a no-win situation that can be avoided if the discloser takes proper precautions. For example, a discloser should never volunteer information which is *outside* the scope of the subject matter as defined in the confidential disclosure agreement. Thus, a lunch break with discloser and disclosee, where martinis are on the menu, should be participated in only by a discloser who either doesn't drink, or by one who is not made too talkative by drinking.

There is an alternative to the potentially troublesome practice of exchanging written summaries days after a meeting: the parties can agree to *summarize* the nature of the disclosure *immediately* following the meeting. Agreement should be reached *at that time* regarding the content of the summary, then such summary can be made a part of the confidential disclosure agreement. This procedure is workable because the parties are acting while the extent of the disclosure is still fresh in their minds. On the other hand, if the parties depart and then subsequently correspond, setting out what was disclosed, problems are much more likely to arise. Perhaps most of such problems result from a failure of memory.

If during a meeting between the discloser and the disclosee it becomes apparent that the subject matter identified in the confidential disclosure agreement is too restrictive, then the parties can "amend" the agreement by delineating the additional subject matter that they wish to discuss; the additional subject matter should be identified in the manner discussed above. The parties can add the additional subject matter to their original agreement, and date and initial the addition in the margin at the point of amendment. This should then allow for a more extended disclosure. However, before discussing previously unanticipated subject matter, the discloser and disclosee should both be prepared by updating their research records as discussed previously.

A final point on a verbal disclosure that a disclosee subsequently attempts to limit by a communication *after* receiving a disclosure: if the disclosee attempts to characterize the disclosure as limited to certain points, and omits other points which the discloser maintains were disclosed, and these other points come within the definition of the subject matter as described in the confidential disclosure agreement, then the disclosee is in a poor legal position. What the disclosee should have done *before* receiving the disclosure, *cannot* be accomplished *after* a disclosure coming within the scope of the subject matter set out in the confidential disclosure agreement has been received. Unless the disclosee can show that the disclosure was *outside* the scope of the *agreed-to* subject, then the disclosee should be held to an obligation of confidentiality concerning the verbal disclosure.

This then highlights the importance of defining the subject matter of the disclosure in the agreement. If the definition is very broad, then the disclosee can be damaged. If the definition is too *narrow*, the discloser can lose. Thus, both parties should carefully consider the scope of the disclosure before the agreement is signed. This enables both parties to realize the benefits of the relationship and not suffer undesirable legal expenses or loss of credibility.

The agreement shown in Figure 4.1 (see page 32) can be used in a situation where an individual wishes to make a confidential disclosure to a corporate entity. The corporate entity is willing to receive the disclosure to determine its interest, if any, in the disclosure. Of course, if there is an interest on the part of the corporate entity, negotiations with the disclosee would then be the next step.

4.1 Modification of the Confidential Disclosure Agreement

The form shown in Figure 4.1 can be modified to contain other recitations which the parties desire. If the disclosure includes the transmission of a subculture of a microorganism which is not publicly available, then conditions pertaining to the disclosee's use of the culture can and should be added. Some of the most important conditions which can be added to a confidential disclosure agreement are as follows:

1. The Disclosee should not subculture the microorganism except where necessary to evaluate its interest in the disclosure.
2. The Disclosee shall not accession the culture into the

Disclosee's general culture collection. (This condition is especially important if the disclosee is a large corporation whose general, or central, culture collection provides cultures for numerous research groups.)

3. Upon completion of its evaluation of the disclosure, the Disclosee shall return all remaining cultures to the Discloser or destroy them.
4. The Disclosee shall use the microorganism for the sole purpose of evaluating its interest in the disclosure.
5. The Disclosee shall not perform any molecular biology work on the microorganism unless such work is necessary for it to evaluate its interest in the disclosure.
6. The Discloser shall not be liable for any injury resulting from the Disclosee's use of the microorganism in evaluating its interest in the disclosure. (This condition does not absolve the Discloser from its responsibility to inform the Disclosee of any information it has concerning the toxicity and/or pathogenicity of the microorganism. Further, the Discloser has a duty to release a microorganism only to those skilled in the use and maintenance of such microorganisms. Thus, before a microorganism is released to another party, the Discloser should ascertain the competency of the Disclosee in this area.)

Researchers should recognize that whenever cultures of microorganisms are sent through the mail or transported on public conveyances, there are a variety of state and federal regulations that need to be followed. These regulations are designed to insure the safety of the public should a container in which a culture is being transported be damaged during transit. Information concerning these regulations is available from a variety of sources, such as the American Society for Microbiology, 1913 I Street, NW, Washington, D.C. 20006 USA, and the American Type Culture Collection, 12301 Parklawn Drive, Rockville, Maryland, 20852 USA.

The procedures described in this chapter are for maintaining the confidentiality of research work when the researcher wishes to disclose the work to a person or entity outside the particular work environment. However, when the researcher wishes to *publish* the research work, then confidentiality is short-lived, and the effects on legal protection can be significant. This topic will be discussed in the next chapter.

5

Publishing Research Work

A prime desire of most research scientists is to publish their research work for peer review. Unfortunately, however, premature publication can have certain adverse effects upon any subsequent attempt to obtain patent protection for the work disclosed in the publication. Furthermore, the possibility of trade secret protection can also be seriously compromised by publication of research work. Therefore, it is very important for scientists to realize the legal consequences of publishing their work.

5.0 What is a Publication?

The term "publication" has a specific legal meaning in the context of patents and trade secrets. In the legal sense, a publication is a conveyance of information in such a way that the

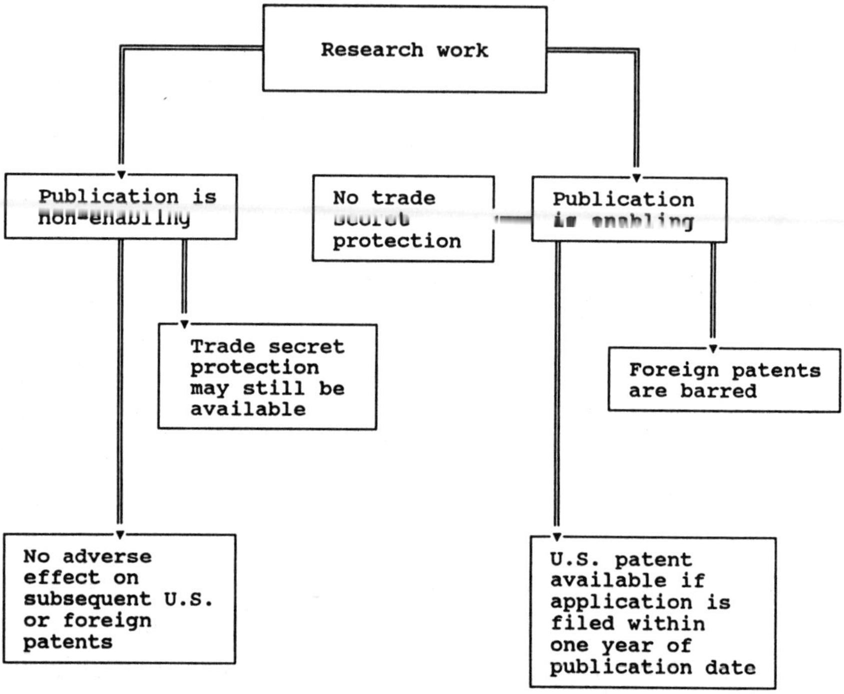

Figure 5.1 The legal consequences of publication depend to a large extent on whether the publication is enabling or non-enabling.

information is potentially available to the general public. Furthermore, in order to have legal significance, the publication must be of sufficient detail to enable a person skilled in the relevant art to practice or duplicate the work. Such a publication is known as an "enabling" disclosure. For example, if the publication pertains to the making of chemical compound X, and the details given are sufficient to enable a chemist to make compound X, then the publication is an enabling disclosure. Because it is an enabling disclosure, it destroys the confidentiality of the subject matter of the disclosure. Once confidentiality is destroyed, it can no longer be recaptured.

Therefore, if publication is being considered, the researcher should first reach a decision regarding the preservation of legal rights.

The effect of an enabling disclosure and the loss of confidentiality on trade secret rights and on patent rights differs. Figure 5.1 illustrates the effect of enabling and non-enabling publications on intellectual property rights.

5.1 Publication—Effect on Trade Secret and Patent Protection

The publication of research work in an enabling manner completely eliminates the use of trade secret protection as a possible legal right. This is because the crux of a trade secret is that the subject matter of the research is *secret*. However, patent protection is still available in the United States if a patent application is filed in the United States Patent and Trademark Office (PTO) within one year of the *publication date*. The publication date is when the publication actually becomes *available* to the public, for example, on a public library shelf or at a newsstand. The date an article is received for publication by a journal is not the date from which the one-year period for filing a patent application begins.

If a patent application is not filed in the U.S. PTO within one year from the date the subject matter was first disclosed to the public in an enabling manner, anywhere in the world, then a patent can never be obtained for the subject matter in the United States. This is referred to in legal parlance as a *"statutory bar,"* or more simply, as a *"bar."* In most foreign countries, there exists the feature of *absolute novelty* in the patent laws. This means that any publication of an invention prior to the filing of a patent application, bars the obtention of a patent in these countries. That is, unlike in the United States, there is no one year grace period. There are certain limited exceptions but, basically, if a publication is made prior to the filing of a patent application, foreign patent rights are lost.

One very important feature of international patent practice is that it is possible for an application to receive the benefit of the filing date of an earlier filed application on the same invention. That is, when an applicant first files a patent application in any country, subsequent applications for the same invention in other countries may have the date of the filing of the original application as their effective filing date. In order for a subsequently filed application to receive the benefit of the filing date of an earlier application, the subsequent application must be filed within one year of the filing date of the original application. For example, if a United States patent application were filed on July 1, 1989, and a Japanese application for the same invention was filed on June 30, 1990, the Japanese application would have an effective filing date of July 1, 1989. However, if the Japanese application were filed on July 2, 1990, then the Japanese application would not receive the benefit of the earlier filed application, and, therefore, the filing date of the Japanese application would be July 2, 1990. The earlier effective filing date can be of critical importance when a publication on the applied-for invention has occurred in the time between the original filing of the application and the subsequent filing. For example, in the scenario just described, where a Japanese application was filed on June 30, 1990, if an enabling publication occurred on October 1, 1989, the applicant would be barred from obtaining a Japanese patent were it not for the effective filing date of July 1, 1989. Figure 5.2 summarizes this information.

Researchers should recognize that some inventions may be patentable in the U.S. but not in various foreign countries. For example, a process for treating the human body is pat-

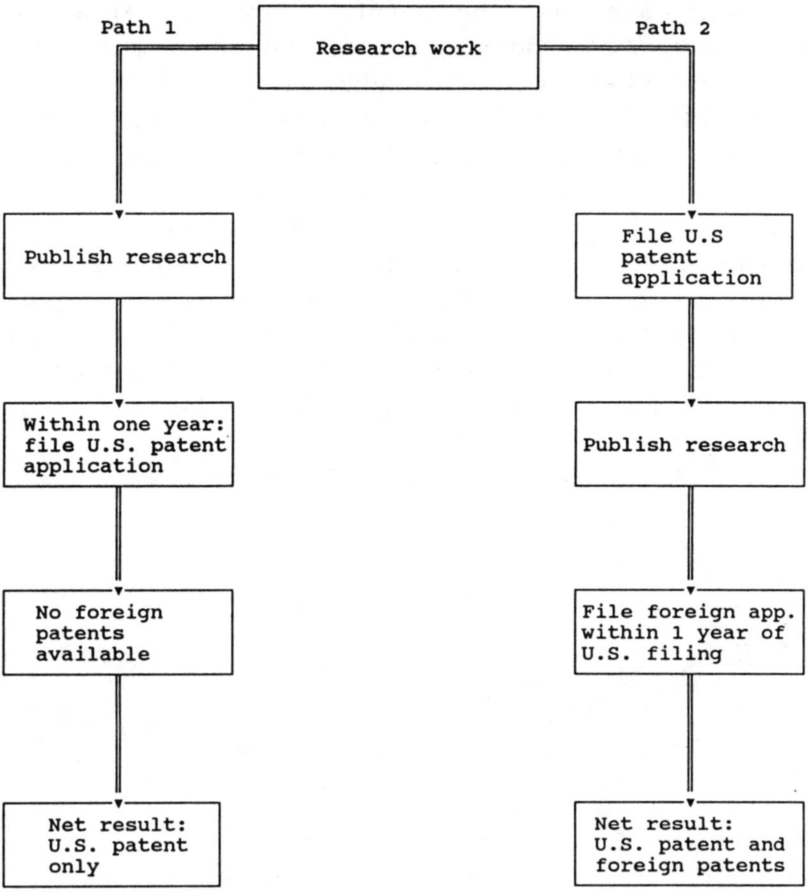

Figure 5.2 Two paths for publishing and patenting: Premature publication can jeopardize both foreign and domestic patent rights.

entable in the U.S., but not patentable in the European Patent Office (EPO). Thus, publishing of the invention prior to filing a patent application may have no adverse effect because a foreign patent could not have been obtained anyway, and a United States patent can still be obtained if an application is filed within one year of the date of the publication. However, a full assessment of domestic and foreign patent possibilities should always be made *before* publication.

Unfortunately, it is not too unusual in the real world to have a scientist publish research work before any patent considerations have been made. A notable example is the publishing of work done by Drs. Stanley Cohen and Herbert Boyer. This work resulted in the issuance of U.S. Patent 4,237,224 on December 2, 1980; this patent covers some of the basic techniques used by many molecular biologists. However, the publication of this work occurred before any patent application was filed. Thus, there are no foreign patents for the Cohen-Boyer invention.

6

Disclosure of Research Work to a Government Agency

Disclosures of research work are made to government agencies for a number of reasons. A researcher may have to make a disclosure of past and proposed work in conjunction with a grant application, or a disclosure of an invention may have to be made to a government agency in conjunction with a request to manufacture or test the invention. There are numerous such instances, but, in each instance, confidential proprietary information is going to an "outsider."

There are general guidelines which can be used in an attempt to protect and preserve the confidentiality and proprietary nature of submissions to a government agency. However, first and foremost, it should be understood that any submission to a government agency should be sufficiently complete and accurate so that the agency can make a proper determination of the issue involved. The researcher should try to ascertain how much information needs to be submitted. This can be done by studying the particular government regulations covering the submission and by communicating with pertinent agency persons.

When a submission is made, material which is confidential or proprietary in nature should be clearly marked. It is the general practice of government agencies in the United States to make every attempt to preserve the confidentiality of the submitter's material. Yet, there can be instances when, for one reason or another, the system fails or disclosure is required by law. In these circumstances, it is to the submitter's advantage to have already filed a patent application in the PTO, assuming the subject matter is patentable. Unfortunately, if the subject matter is only suitable for trade secret protection, then a submitter's trade secret could be in danger of being lost.

It is not likely that a government agency will mishandle a confidential disclosure. Thus, this potential risk versus the expected benefit from the giving of a grant, or regulatory approval, is an acceptable risk in most instances. Certainly, it is an acceptable risk if the submitter *must* gain agency approval to further test or manufacture an invention. In such a situation, it is strongly recommended, however, that patent action be taken prior to the submission to the government agency; assuming, again, that the invention is considered to be a patentable invention.

In giving a *sufficient disclosure* to a government agency, the submitter will often be able to comply with the regulations while retaining in confidence vital parts of the invention. So long as the government agency accepts the disclosure as sufficient, and the submitter has not misled the agency, or omitted relevant material which the submitter knows the agency should have, the submitter can retain certain confidential elements without disclosure. It should be recognized that this is not intended as a suggestion to "play games" with a government agency. Any attempt to do so will ultimately work to the disadvantage of a submitter.

In an analogous situation with regard to submitting to a government agency, in this case to the PTO, there are court decisions which hold that a patent applicant is not required to disclose *production* data in a patent application for a new process. For example, the invention may be a process for making an antibiotic using a better strain of a microorganism, or different fermentation medium ingredients. A sufficient disclosure for patent purposes needs only to include shake flask or pilot plant data to support the patentability of the process. There would be no need to disclose production data, such as fermentor size, fermentation titers, and overall production yield data. To a scientist involved in antibiotic production, this information is confidential and every attempt is made to keep it so. However, if the invention can only be shown to occur in the production system, then, of course, a sufficient disclosure would require the disclosure of the production system. This type of situation is rare because most process inventions at the bench or pilot plant stage can be scaled up to a production level even though the magnitude of the advantage at this level may be less than that of the developmental stage. For example, a microbiological process in a shake flask or a pilot plant fermentor may give a titer increase of 30% over control whereas the scaled-up process may give only a 20% increase. Still, this increase would be significant and would not preclude a finding of patentability.

Even if a government agency requires manufacturing data as a part of its regulatory review process, fermentation titers and overall yield data may not need to be included. Generally, the government agency is primarily interested in the nature of the equipment, facilities, procedures, and materials used to manufacture a product. Actual product yield is not ordinarily considered to be relevant or necessary data.

The overall tone of submissions to government agencies should reflect a total spirit of compliance; however, informed judgment should be exercised as to what constitutes a full and

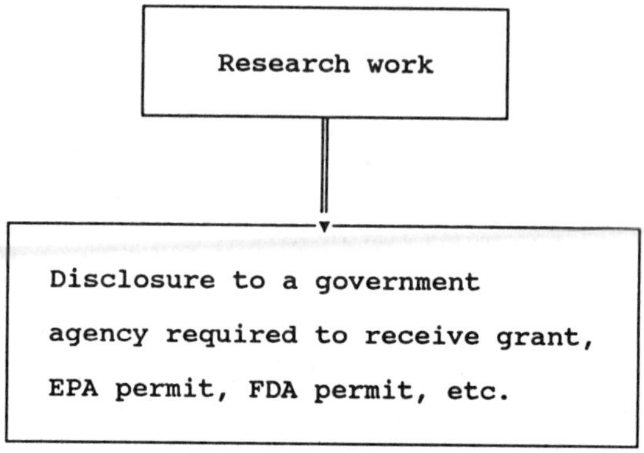

1. Explicitly identify confidential or trade secret parts of disclosure.
2. Give a full disclosure to allow agency to make the proper determination.
3. Do not disclose production yield data that is irrelevant to the agency consideration.

Figure 6.1 The proper procedure for disclosure to a government agency.

sufficient disclosure. Any questions arising in a particular situation might best be addressed by communicating with the government agency before making a submission. Finally, if patent action is possible, this should be taken *before* any submission is made. Figure 6.1 summarizes the information given in this chapter.

7

The Effects of Prior Public Use

In addition to publication, the other major route by which the confidentiality of inventions can be lost is by "public use." The term "public use," like the term "publication," has a specific legal meaning. A public use occurs when an invention is used in such a way that the general public is not prevented from observing the use.

The public use of an invention can have important legal ramifications. For example, under United States law, if a complete invention is in public use for more than one year, there arises a "statutory bar" to the subsequent obtention of a patent for the same invention, that is, a patent can never be issued for this invention. This rule was formulated under the theory that inventions that are in the public domain for more than one year should not thereafter be withdrawn from public use by the issuance of a patent. Usually, foreign patent rights remain available indefinitely, so long as the public use is not in a country where a patent is sought. Figure 7.1 summarizes the

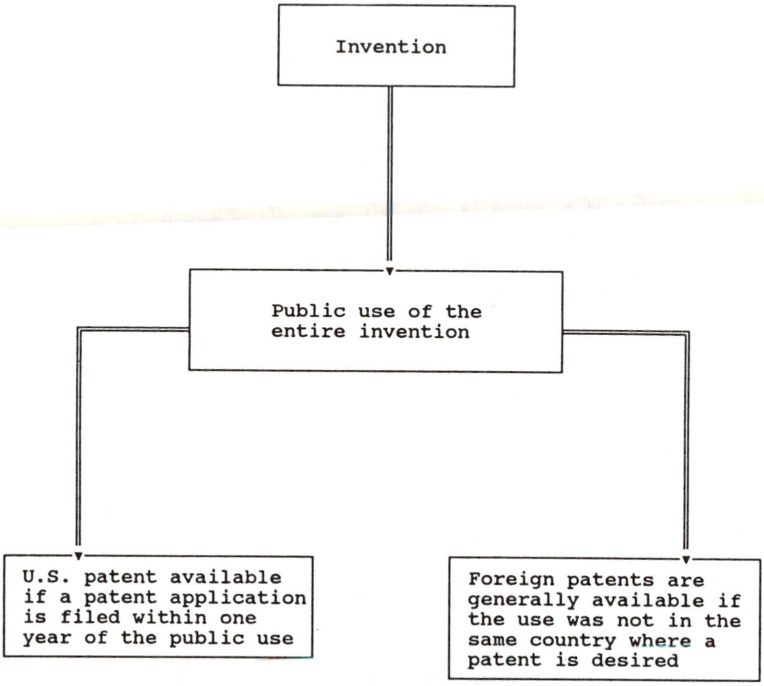

Figure 7.1 The effect of prior public use of an invention on the possibility of obtaining U.S. and foreign patent protection.

difference between U.S. and foreign patent rights in this respect.

There are various exceptions and conditions to the general rules pertaining to prior public use. For example, not all public uses have the same legal ramifications. In particular, a public use which is necessary to evaluate and perfect an invention is referred to as an "experimental use." An experimental public use has different legal consequences than a public use of a perfected invention. Because both of these terms and the events that they refer to are of premier importance in the field of biotechnology, it is very important that researchers understand them and their implications.

In the United States, an inventor has one year from the date a perfected invention is used in the public in which to file a patent application for that invention. If the invention used in the public is an incomplete embodiment of the invention, then the clock will not start to run until the *complete* invention is used in the public. For example, if the complete invention consists of ingredients A, B, C, and D, then a prior public use of only ingredients A and B will not start the one year clock for filing a U.S. patent application.

Cases occur frequently in the biotechnology area where a particular invention may seem to be essentially complete, but uncertainty remains as to whether the invention will be suitable for the purposes intended. Initially, inventions are tested in the laboratory, but success there does not necessarily mean that the invention will perform properly in actual use. Therefore, inventors will frequently go through an experimental period to test the invention under contemplated use conditions. Such testing would be considered to be experimental use. This test period generally will not cause the statutory one year period to start running until the test results prove positive. This is illustrated in Figure 7.2.

An example of experimental testing in the biotechnology area is the testing of a bioherbicide in the field. Even when a particular bioherbicide is effective in greenhouse tests, there may remain uncertainty as to whether it will be effective in the field under contemplated use conditions. Therefore, when a bioherbicide undergoes such field testing, it is advisable to clearly state on the testing protocols that it is an "experimental test." In fact, governmental experimental use permits for field testing require such notification. It is only after such field tests are completed, and the bioherbicide has been shown to be effective, that the statutory one year period will begin to run if further public use is made. One example of further public use is the practice of distributing samples of the bioherbicide to potential customers in order to determine customer ac-

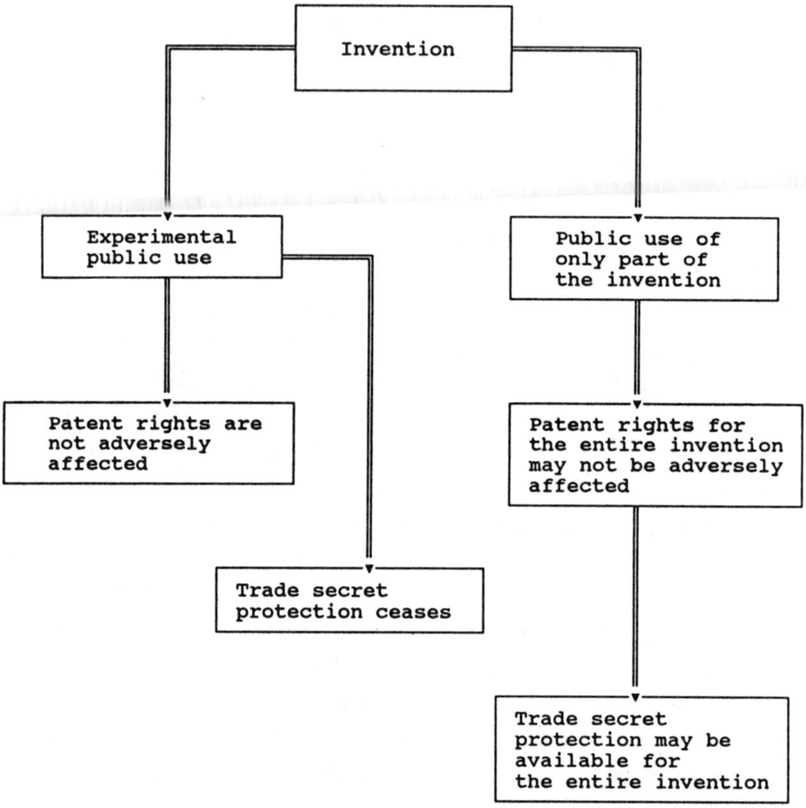

Figure 7.2 The effect of experimental public use or of public use of only part of an invention on obtaining patent and trade secret rights.

ceptance. Also, efforts directed to marketing the bioherbicide *after* the effectiveness of the bioherbicide has been established under contemplated use conditions, do not constitute experimental use. Therefore, such commercializing efforts do not prevent the running of the statutory one year bar.

Before leaving this topic, the reader should be alerted to the fact that the term "experimental use" is also used in a completely different patent law context. This other usage of the term "experimental use" has relevance in *infringement* determinations. In general, a patentee has the exclusive right to make, use, and sell his or her invention. Use by an individual other than the patentee or licensees constitutes infringement. However, one exception to this general rule is that a use which is for experimental, non-commercial, purposes does not constitute an infringement. A complete discussion of this "experimental use doctrine" is beyond the scope of this book. However, one recent piece of legislation relating to this type of experimental use is of particular importance to the biotechnology industry. In 1984, Congress passed the Drug Price Competition and Patent Term Restoration Act which, among other things, established that the production of patented drugs in sufficient quantities to satisfy U.S.D.A. drug registration requirements can be considered an experimental use and, thus, does not constitute infringement. The primary purpose of this digression is merely to alert the reader that the term "experimental use" is applied to two different situations relating to patent law.

8

Choosing Trade Secret or Patent Protection

Assuming that confidentiality has been maintained by using the proper procedures during the research and development phase of a research project, at some point it is necessary to make a decision as to whether trade secret or patent protection is the best form of legal protection for the research work. As discussed previously, trade secret requirements must always be kept in mind to avoid losing confidentiality to an outside party. If all the trade secret criteria are still in place, the researcher has a choice between continued use of trade secret protection or pursuing patent protection. This option initially arises when an invention is made and the information needed for filing a patent application becomes available.

In order to determine the patentability of an invention, it is necessary to be cognizant of the prior art to which the subject matter of the invention pertains, as well as of the requirements for patentability as set out by the patent law and as interpreted in various legal decisions. Obviously, if the researcher wishes to patent an invention in the United States, he or she must turn to someone who understands U.S. patent law. On the other hand, patenting in other countries requires knowledge of the patent laws of the particular country where a patent is desired.

It should be evident that making judgments about trade secret or patent protection is properly reserved for the trained professional, i.e., a patent lawyer. However, the patent lawyer must work in concert with the scientist in order to make a proper judgment. The scientist is in a vital position to give positive input in this decision-making process. Accordingly, it is advantageous for the scientist to have some understanding of the factors impacting on this decision-making process. The next section contains a discussion of the factors employed to determine whether trade secret protection or patent protection is the best form of protection for a particular invention.

8.0 When Trade Secret Protection is the Only Alternative

If an invention is found *not* to satisfy the requirements for patentability, then trade secret protection is the only alternative (see Figure 8.1). There can be a variety of reasons why a particular invention may be judged to be not patentable. For example, the invention may be merely a technical adjustment over what is already known to the public, and, therefore, *obvious* in light of the prior art. This judgment of obviousness should never be made by anyone other than a patent lawyer, because extensive training and experience are essential to reach the proper conclusion. Even the most highly skilled patent lawyer will err on this determination at times and an untrained

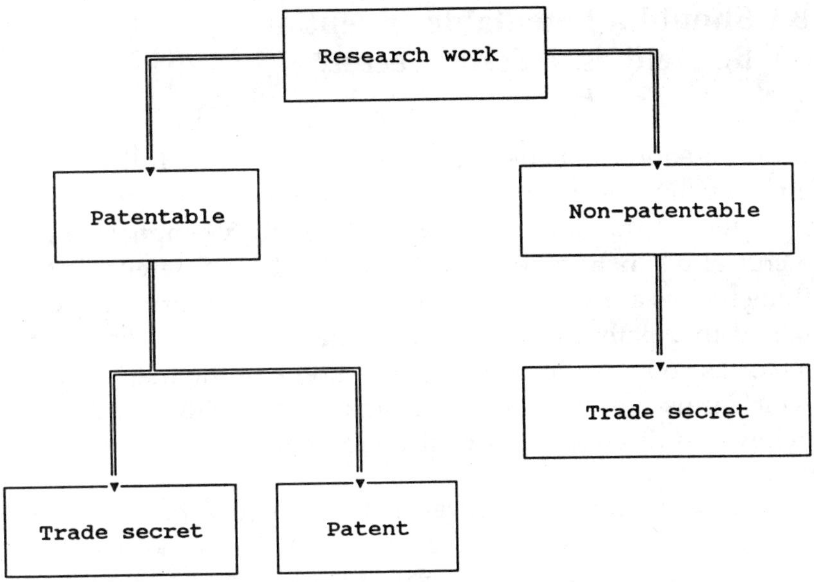

Figure 8.1 The types of legal protection available for patentable and for non-patentable inventions.

person is certainly more likely to make the wrong decision. In fact, skilled patent lawyers will generally be very cautious about making a judgment that a particular invention is not patentable because it is obvious. Some patent lawyers have been haunted by their determination of obviousness when a patent on the same invention was subsequently issued to another person.

Of course, obviousness is but one factor which can render an invention unpatentable. There are also many other points which must be considered, and all require an intimate knowledge of the pertinent field of science and of patent law. However, once an informed decision is made that an invention is not patentable, trade secret protection becomes the only effective means of legally protecting the subject matter.

8.1 Should a Patentable Invention Be Held as a Trade Secret?

If an invention is believed to be patentable, then the patent attorney must decide whether trade secret or patent protection is better. At this point it is worthwhile to re-emphasize that a crucial distinction between trade secrets and patents is that trade secrets are kept in *confidence*, while the information contained in a patent document is made *available to the general public* as soon as the patent issues. Some of the major factors which must be considered in making this choice are listed below and discussed in detail thereafter.

1. Can any patent covering the invention be adequately enforced, i.e., would the patent holder know when the patent was being infringed, and by whom?
2. Can any patent covering the invention be easily circumvented by someone, i.e., could a person vary the invention conditions somewhat so as to avoid the patent claims and still benefit from the invention disclosure?
3. Is the "life" of the patented invention relatively short, i.e., three to five years; for example, an invention in a rapidly moving field will have a short life; or can the life be expected to run the full 17-year term of the patent?
4. Does the disclosure of the invention in the patent document allow persons in the public to greatly accelerate the pace of their research work, perhaps to the ultimate detriment of the patent owner?
5. Is the inventor intent on publishing the invention as soon as possible in order to be considered the "first" to do some important work, such as cloning a particular mammalian gene into a bacterium?

6. Will it be difficult to continue maintaining the secrecy of the invention if the inventor decides to maintain it as a trade secret?
7. Will the financial cost of obtaining patent protection be too high?
8. Are there other scientists active in the particular research area and could another person be reasonably expected to make the same invention?

Factor No. 1: Can the Invention be Reasonably Policed? The value of a patent will depend, in part, on the ability of the patent owner to prevent others from using the claimed invention without a license from the patent owner. If a person could use the invention without the patent owner being aware of such use, then a patent is not an effective means for protecting the invention.

The fact that patented inventions are used by others without prior license from the patent owner should not surprise anyone. Some of the reasons for doing so are as follows: the users believe the patent is unenforceable in a court of law, they believe the patent owner will never learn of their use, or they may not know that a patent has been issued for the invention.

Those who use a biotechnology invention without the consent of the patent owner can be in any of the above groups. However, because of the nature of many biotechnology inventions, the group which believes that the patent owner will never learn of their use is perhaps the largest. For example, a patented microbiological process using a new fermentation ingredient could be used without such use being evident from the resulting marketed product. Thus, the patent owner would not readily know of this infringing use. Such inventions are generally referred to as being difficult to "police." It should be known, however, that once it is learned by the patent owner that such an infringing use has been occurring, it is possible to obtain multiple financial awards from the infringing user in a court action. Thus, a person who knowingly uses a patented process without the consent of the patent owner may be required to pay triple damages as a penalty for his or her infractions. The triple damages penalty, which frequently amounts to millions of dollars, is intended to discourage intentional infringement. Unfortunately, however, infringement of inventions which are difficult to police is still a significant problem. Therefore, if it would be difficult for a patent owner to detect infringement, the invention might best be protected as a trade secret.

Factor No. 2: Can the Patent Covering the Invention be Easily Circumvented? If the protection which could be obtained through a patent on a particular invention could be easily circumvented or "invented around," then trade secret protection may be the best for that invention. This factor can best be understood with the help of a brief background concerning the law of patent infringement. As previously mentioned, a United States patent gives the patent holder the exclusive right to make, use, or sell his or her invention for a period of seventeen years from the date of issuance of the patent. As a general rule, patent infringement occurs when another person makes, uses, or sells the patented invention without the permission of the patent holder. In legal parlance, infringement may be either "literal" or under the "doctrine of equivalents." Literal infringement occurs when an unauthorized person uses the exact invention which is claimed in the patent. Infringement occurs under the doctrine of equivalents when an unauthorized person uses a device which performs substantially the same function as the patented invention and in substantially the same way. Therefore, a patent holder can exclude others from using the exact invention claimed in his or her patent and any other "equivalent" embodiment of that invention.

Some inventions have a wide range of equivalents, while others have a rather narrow range of equivalents. The range of equivalents for any particular invention depends largely upon the subject matter of that invention and the way in which the law pertaining to that subject matter has developed. Clearly, the patent protection for an invention with a narrow scope of equivalents is much easier to circumvent.

For example, if the patented invention covers specific multiple-site cloning vectors, then a person skilled in the art may be able to alter the vectors in such a way that there is no *literal* infringement of the patent claims, and the altered vectors can be shown *not* to be equivalent to the claimed vectors. Under these circumstances, there is no infringement of the patented invention.

Patent disclosures are supposed to foster further developments and inventions. Also, inventing around, or circumventing, a patent claim is a legitimate activity that is fully within the patent law. Accordingly, the inventor and his or her lawyer must carefully assess the potential for such circumvention with regard to the invention at issue. Thus, if the invention would be relatively easy to circumvent, the inventor should think seriously about maintaining the invention as a trade secret, rather than patenting it.

Factor No. 3: Is the Life of the Patented Invention Relatively Short? The expected life of an invention varies greatly depending on the particular art to which it pertains. In certain areas of biotechnology, new advances are made almost daily. This means that many biotechnology inventions become obsolete quickly, that is, they are relatively perishable. Patenting such a "perishable invention" is generally not worth the financial cost, unless the inventor simply wishes to have a patent to his or her credit. This is so because it could be expected that new advances will make the perishable invention obsolete within a short time. It should be recognized, however, that even in fast-moving research areas, *pioneer* inventions are made. A pioneer invention is one that is the basis for many other subsequent inventions. These pioneer inventions are often not perishable, and, further, are not easily circumvented. Such pioneer inventions are best protected by patenting if all the requirements for patentability are met. The Cohen-Boyer U.S. Patent 4,237,224 for basic techniques of genetic engineering is an example of an invention which may turn out to be a pioneer invention.

Factor No. 4: Does the Patent Disclosure Give Competitors an Edge? The disclosures contained in patent documents are tremendous sources of information that are available to all the public. In the past, scientists have relied primarily on scientific articles in order to keep abreast of recent developments in their field of research. However, more and more scientists are recognizing that it is very worthwhile to examine the content of newly issued patents.

By law, patents *must* contain a complete, enabling disclosure of the claimed invention. Such a complete disclosure is not always found in scientific articles. The enabling disclosure of a patent is the *quid pro quo* required for the patent grant. The information contained in a patent disclosure may encourage other scientists to pursue a course of research that they would otherwise not have undertaken. This push in the right direction could lead to other inventions, which, ultimately, might compete with the patented invention. Of course, this is one of the purposes of the patent system, and it works to the overall benefit of society. However, the individual inventor must assess the possibility that his or her patent disclosure will give other researchers the ability to conduct research which might eventually produce a product that would compete with, or replace, the patented invention. The inventor must keep in mind that, once the patent route has been chosen, a full and enabling disclosure of the invention must be given.

It should be noted at this point that the *filing* of a patent application in the United States PTO does *not* destroy the secrecy aspect of a trade secret. This is because pending patent applications are, by law, maintained in secret. Also, the inventor has the option of "abandoning" the application at any time up to issuance of the patent, thus retaining secrecy.

Factor No. 5: Does the Inventor Want to Publish the Invention? A frequently encountered situation, especially with inventors employed in academia, is a desire to publish the invention as soon as possible. It should be remembered that an enabling publication of an invention destroys the possibility of maintaining the invention as a trade secret. To be an enabling publication, the invention must be disclosed in sufficient detail to enable a person skilled in the art to reproduce the invention. This ability to reproduce the invention need not be exclusively from the publication itself, but, rather, the publication in combination with what is already known in the art.

It is quite possible in the biotechnology art to publish an invention without disclosing a key element of the invention which is essential to reproduce the invention. For example, if the invention concerns a new antibiotic which is made using a new microorganism, a publication would be enabling only if it disclosed how to make the antibiotic via chemical synthesis, or disclosed *and* made available the novel microorganism necessary to make the antibiotic by fermentation means. A new antibiotic is generally characterized initially by various chemical and physical parameters such as infrared absorption spectrum, elemental analysis, ultraviolet absorption spectrum, optical rotation, or melting point. This kind of characterization, even though sufficient to identify the new antibiotic and distinguish it from other antibiotics, is not enough to allow for chemical synthesis. Even if the structural formula of the antibiotic is given, this in most instances would not enable a person skilled in the art to make the antibiotic.

The microbiological process for making the new antibiotic would be the essential information that a scientist would need to make the antibiotic. Therefore, if the novel microorganism is not disclosed in the publication in such a manner as to allow a person to obtain a culture of the microorganism, then the publication would not be an enabling disclosure. Generally, when an invention requires the use of a microorganism, the only way to make a publication an enabling disclosure is to disclose a source from which a culture of the microorganism could be obtained. This is usually done by disclosing the deposit accession number given to the culture by the depository where the culture is on deposit and from which it can be obtained (see Chapter 10 and Appendix B). If an accession number is given, but a person in the public cannot obtain a culture from the disclosed depository, then it could hardly be argued that the publication is an enabling disclosure. However, if a culture is available, even if a fee is charged, the publication may be considered an enabling disclosure.

It should be noted that in keeping with the spirit and norms of scientific research, as well as the policy of many major journals, it is generally not acceptable to withhold information which is necessary in order to verify research results. Therefore, it cannot be recommended that scientists attempt to retain trade secret protection by publishing nonenabling scientific articles.

Therefore, those who wish to publish speedily must recognize that they are probably giving up their trade secret protection. Furthermore, if they publish before filing a patent application on the invention, they are placing in jeopardy significant foreign patent rights, as was discussed in Chapter 5. Though the United States patent law gives a person a grace period of one year from the date of publication in which to file a United States patent application for the published invention, most foreign countries do not provide this leeway. Thus, if an invention may be important world-wide, it would be foolish to publish the invention before first taking patent action.

In summary, it should be recognized that a decision to publish an invention is a decision to forego trade secret protection and jeopardize patent protection. Accordingly, all the ramifications of such a decision should be thought out before the actual publishing.

Factor No. 6: Will it be Difficult to Maintain the Trade Secret? Many trade secrets are difficult to *maintain* as trade secrets for one reason or another. This, then, *must* be considered when going through the procedure of evaluating whether to choose trade secret protection over patent protection. If the trade secret can be confined to a small number of persons in a "closed door" type of environment, then maintenance of the trade secret is not difficult. For example, maintaining a novel yeast culture in secret for a number of years has been possible in many brewery operations. Also, high-producing strains of antibiotic-producing microorganisms are routinely held in secret by most antibiotic manufacturers. However, even despite stringent precautions, there are still known instances where valuable antibiotic-producing strains have been stolen, or at least transferred without authority, to outside persons for financial remuneration. The case of the transfer of tetracycline strains to foreign manufacturers in the early stages of the antibiotic era is one example. However, in such instances, as was the case in the tetracycline episode, the perpetrators are often apprehended and prosecuted for theft of valuable property. This possibility of theft or conversion of trade secrets is a factor that must always be reckoned with. Extra precautions should be instituted if a trade secret is particularly valuable and there is a known market demand.

Generally, maintenance of select strains of microorganisms as trade secrets is not a difficult problem. On the other hand, maintenance of a trade secret in an area accessible to outside persons such as repair persons and sales persons can be a difficult task. Where such an environment exists, then it becomes incumbent upon the trade secret owner to take extra steps to ensure the continued secrecy of the trade secret. If the extra steps are not worth the time and expense, then trade secret protection may not be the desired course of action.

Factor No. 7: Is the Cost of Patenting too High Relative to Maintaining a Trade Secret? The cost of maintaining a trade secret must be balanced against the cost of patenting the invention. Generally, the cost of maintaining a trade secret, as measured by "out-of-pocket expenses," is not as high as that of obtaining a patent. However, if the trade secret may be difficult to maintain in secret without taking extra precautions, then patenting costs may be less.

Factor No. 8: Is the Invention in an Active Area of Research? If other scientists are actively working in the area of the art to which the invention pertains, rapid filing of a patent application is often the best course of action. This is especially true if it can be reasonably expected that others would file patent applications for their similar inventions as soon as they are made. This is because, under U.S. patent law, the first person to file a patent application has certain procedural advantages if an interference proceeding ensues.

Also, though the United States patent law allows for the determination of the first inventor by proof of notebook invention events, foreign patent laws award the patent to the first person who filed a patent application for the invention. Accordingly, where the research area is "crowded," rapid patent action is the desired course of action.

Another factor which favors rapid filing of patent applications is that if a United States patent is issued for an invention that another person has been maintaining as a trade secret, the trade secret owner will no longer be able to practice the trade secret without permission from the patent owner. This is so even though the trade secret owner has been practicing the trade secret for years prior to the issuance of the patent. And, as mentioned in Chapter 7, the trade secret owner cannot patent the invention held as a trade secret if the invention was used commercially for more than one year. This prohibition to patent on the trade secret owner does not spill over to another person who makes the invention independently of the trade secret owner.

The above illustrates the encouragement, in effect, that is found in United States patent law to seriously think about patenting an invention, rather than maintaining it as a trade secret.

Laws in certain foreign countries, such as Germany, allow a prior trade secret owner to continue *personally* practicing a trade secret, even after a patent covering the invention of the trade secret is issued in Germany. There are those who have argued for a similar provision in the United States law. However, those not favoring a change in the United States law maintain that the present United States law encourages inventors to seek patent protection. Thus, the law favors a strong patent system, which in the long run should result in greater scientific achievements.

As a practical matter, this bar to continued practice of a trade secret after a patent issues to another is often critical in chemical and fermentation process industries. Often, changes in process conditions or reagents which lead to a more profitable process are held as trade secrets. This is because there is a reluctance to expose these changes to others via a patent disclosure. Such a decision not to patent requires consideration of the competitive position of the trade secret manufacturer should someone else patent the invention of the trade secret. If the trade secret is an absolute necessity for maintaining competitiveness, then it may be reckless to maintain it as a trade secret, if it could reasonably be expected that a subsequent patent owner would not be willing to grant a license to the trade secret owner under a patent covering the invention.

Nevertheless, there are situations where trade secrets are validly maintained over many years, much to the advantage of the trade secret owner. The formulas for Coca-Cola® and Listerine® are two good examples of successful trade secrets. Such well-known and successful products have been copied but, apparently, never faithfully duplicated. Thus, the true products retain their great market share. Perhaps the sophistication of the times and the ability to reverse-engineer products may not allow such long-term trade secrets to exist in the future. If this is the case, then a strong patent system will become even more important as a means for protecting intellectual property rights.

8.2 Summary of the Advantages of Trade Secret Protection

Having discussed some of the factors to be considered in the process of deciding whether to maintain a trade secret, the following summary of the advantages and disadvantages of trade secrets, and of affirmative acts necessary for the maintenance of trade secrets, may be useful.

The *advantages* of trade secret protection can be summarized as follows:

a. The expenses of obtaining patent protection and enforcing patent rights are not encountered when trade secret protection is used. There *are* expenses in the maintenance and enforcement of trade secret rights, but such expenses are generally less than patent costs.

b. There is no time limitation for trade secret protection; the only limitation is that secrecy must exist. It is theoretically possible to maintain trade secret protection for many years more than the 17-year term granted with a United States patent. Realistically, the term of a trade secret in a fast-moving area of science may be even less than 17 years.
c. Competitors are not apprised of the trade secret, as contrasted to the full disclosure required for a patented invention.
d. In many microbiological and genetic engineering inventions, competitors are not able to practice the trade secret invention without a specific microbe or clone. Therefore, it is possible to disclose to the public some elements of an invention without destroying the trade secret protection. In contrast, the filing of a patent application for a microbiological process necessitates the deposit of a subculture of the key microbe in a suitable culture depository (see Chapter 10). The microbe then becomes available to the public upon the grant of a United States patent.
e. A trade secret does not have to be a patentable invention; it just must be unique and secret.

8.3 Summary of the Disadvantages of Trade Secret Protection

The *disadvantages* of trade secret protection can be summarized as follows:

a. There can be no publication of the trade secret in an enabling form. Publication destroys secrecy, unless

Table 8.1 Factors Favoring Trade Secret Protection
1. Invention is considered to be non-patentable.
2. If patented, infringement of the patent would be difficult to detect.
3. If patented, the disclosure of the invention via the patent document would enable a person to easily circumvent the claimed invention.
4. The invention is in a rapidly moving technology where new developments can be expected in a short time, hence rendering the invention obsolete in a short time.
5. The financial cost of obtaining a patent, i.e., patent application preparation and processing fees, is too high for the inventor to bear.

the publication is non-enabling, that is, unless a person skilled in the art could not practice the trade secret from the disclosure in the publication coupled with what is already known in the art, or readily ascertainable from what is known in the art.

b. The trade secret may be independently developed by another inventor who then may obtain a patent. In the United States, the patentee can then enforce the patent against the prior trade secret holder.

c. The passing of a trade secret to a competitor by theft or by breach of a confidential relationship may be done in a manner not easily susceptible to discovery or proof.

d. Reverse engineering of a product presented to the public can be used legitimately to uncover a trade secret. That is, the public can analyze the product to determine its composition.

e. Trade secret protection exists only so long as secrecy exists.

Table 8.2 Factors Favoring Patent Protection

1. If patented, any infringement of the patent would be readily detected.
2. If patented, the breadth of patent protection would be broad enough to give meaningful protection of the invention.
3. The inventor has a desire to publish the invention and, thus, obtain peer recognition.
4. The invention is considered to be a basic invention which could be expected to "dominate" many inventions yet to be made. For example, the Cohen-Boyer patent (U.S. 4,237,224) is a basic patent.
5. Many researchers are active in the invention area and it could therefore be reasonably expected that someone else may make the same invention.
6. Maintenance of the invention as a trade secret is considered to be difficult. For example, if a number of persons need to know the trade secret in order to carry on their jobs, it becomes more likely that a person in the group will breach the agreement of secrecy.

Having summarized the advantages and disadvantages of trade secret protection, it is also helpful to compare the key features of trade secret protection as opposed to patent protection. This is done in the following list. In addition, Tables 8.1 and 8.2 summarize the factors favoring trade secret and patent protection, respectively.

Assuming that the trade secret is a process to make a product which is sold to the public:

1. Trade secret protection continues so long as the process remains a secret. If the process can be ascertained by "reverse engineering," or from the nature of the product sold, the trade secret can be lost.
2. If the trade secret remains secret, and the secret process is used to make the product which is sold com-

mercially for more than one year, patent protection in the United States is no longer available to the trade secret holder. If a patent application is filed in the United States *before* the one-year period expires, then a patent is *not* barred.

3. The filing of a patent application in the United States does not destroy the secret nature of a trade secret since pending U.S. patent applications are maintained in secret, i.e., they are available only to pertinent PTO persons charged with the task of examining the application for patentability.
4. The granting of a U.S. patent will destroy the secret nature of a trade secret since a U.S. patent is published upon issuance.
5. If a patent application has been filed, but a patent has not yet issued, but has been allowed, and the patent applicant decides to continue the trade secret, the patent applicant can abandon the patent application by giving suitable notice to the PTO. An abandoned patent application is *not* available to the public unless it is referred to in an issued U.S. patent.
6. If, after filing a U.S. patent application, foreign applications are filed, for example, in the European Patent Office (EPO) or in Japan within one year of filing the U.S. patent application, the foreign applications will be published approximately 18 months from the U.S. patent application filing date. This publication of the foreign application will destroy the trade secret, even if a U.S. patent has not been granted.
7. If, after filing a U.S. patent application, foreign applications are filed, for example, in the EPO or in Japan, more than one year after the filing of the U.S. patent application, the foreign applications will be published approximately 18 months from the filing date of the foreign applications. This publication of the foreign application will destroy the trade secret.

8.4 Maintenance Requirements of Trade Secret Protection

Maintenance of a trade secret requires certain affirmative acts by the trade secret owner if subsequent attempts to legally enforce the trade secret against misappropriators are to be successful. Some of these affirmative acts are as follows:

a. Public access to any area where trade secrets are being practiced, discussed, or maintained should not be allowed, or at least should be strictly controlled. These areas can include research laboratories, libraries, lounges, cafeterias, and production facilities in a business organization. Should public access be provided to some of these areas, employees who know the trade secret should be instructed not to discuss the trade secret openly among themselves in such areas.
b. Employees should be instructed not to display documents disclosing the trade secret when the employee is in a public area, or even when riding the company elevator.
c. An employee with knowledge of a trade secret must be instructed as to which other employees also have that knowledge so that the employee will not accidentally divulge the trade secret to others.
d. The trade secret should be disclosed only to those who need to know and only under confidential conditions.
e. The trade secret should always be identified as being secret. Documents must be labeled "secret" and employees must be apprised of the secret nature of the trade secret.

9

Patenting the Invention

9.0 When a Patent is the Best Form of Protection

Even before a decision is made to patent an invention, it is advisable to have a patent attorney available who can provide guidance with respect to the best means for protecting property rights and maintaining confidentiality. As previously mentioned, a patent attorney should be primarily responsible for deciding whether to pursue patent protection. Once a decision is made to go ahead with a patent application, then participation by a patent attorney is virtually indispensable.

The matter of selecting a suitable patent attorney is usually quite straightforward for an inventor who works for a large corporation or university. These institutions almost always have a patent attorney on their staff whose job is to advise employees on all matters relating to intellectual property. In addition, corporations and universities may have a

<u>Invention Record and Report</u>

(Submit to the dean of your school or college.)

____
Name of person filing report   Date

1. Brief description of invention:

2. Full name of researcher(s), home address(es), and appointment title(s):

3. Recommendation of researcher(s) whether patent should be sought:____Yes ____No

4. Results to be achieved by the practice of this invention:

5. Outline of means discovered for achieving above results in terms of:
(a) the steps in a process, or (b) the components in a composition or groups in a chemical compound (include description of process of making) or (c) elements in a machine, article, or device. Point out means which are essential, others which are important or useful, and any <u>critical</u> limitations on any of these.

6. Chronology of principal events in conception and developments:

(a) Earliest conception date (reference to substantiating evidence desirable): ______________________

(b) Date of disclosure (orally or in writing) to other persons and names of such persons: ______________________

(c) First written record pertinent to invention: _____

Figure 9.1 The invention disclosure form used at one large university.

(d) Date and result of first test of the invention. If invention is a process, its first test is the first successful trial; if a composition of matter or a compound or a machine, article or device, its first test is its first creation and evaluation with respect to new or improved properties or behavior.

7. Source(s) and amount(s) of <u>all</u> grant, contract or gift funds used by researcher(s) regardless of purpose or use during the period starting with the date noted in item 6(a) and continuing to the present.

8. Identify those sources indicated in item 7 which contributed to the invention.

9. Date and place (e.g., particular periodical) of any publication regarding invention (whether publication has occurred or is projected):

10. Background of published information and practice in the field of the invention (known practices, periodical citations, patents, etc. Attach documents if pertinent.).

11. Features embodied in this invention which would not have been obvious to or readily foreseeable by the typical skilled worker in the field:

____
Signature of person filing report

____
Researcher's signature   Date

____
Witness to researcher's signature

Certification by researcher's supervisor
(department chairman, program director, or coordinator)

I have reviewed the information provided above with particular reference to item 8, source of funds contributing to the invention. To the best of my knowledge, I believe the above statements to be accurate.

____
Signature of supervisor

booklet explaining the institution's policies and procedures with regard to obtaining legal protection for inventions. Also, these institutions frequently provide a form on which inventors supply information concerning their new idea. An example of such a form is shown in Figure 9.1.

On the other hand, if the inventor is not on the staff of an institution that has its own patent attorney, he or she will need to retain a patent attorney. Before engaging the services of a patent attorney, it is desirable to ask the attorney to give information about the type of inventions that he or she has previously handled. This information will enable the inventor to determine whether the attorney could be expected to understand the invention. The ability to fully understand an invention will aid the attorney greatly to do the best job possible in patenting the invention. It is generally not recommended that an attorney with schooling and work experience only in the mechanical arts be retained to prepare a patent application for an invention that requires an understanding of chemistry and biology. Similarly, an attorney with prior training and experience only in biotechnology would not be recommended for handling an invention in the electrical art.

Generally, patent attorneys will not accept the responsibility of handling an invention which is outside the scope of their expertise. However, there are gray areas and it is to the advantage of the inventor to engage the best available services at the most reasonable cost.

The professional fee charged by patent attorneys is often a reflection of the demand for their services. This is no different than, for example, the professional fees charged by various medical specialists. If the invention is in a highly complex field, such as many areas of biotechnology, then the fees can be expected to be higher than fees charged for servicing low-technology inventions.

Professional attorney fees also vary with the location of the patent attorney. Again, this is similar to the practice of

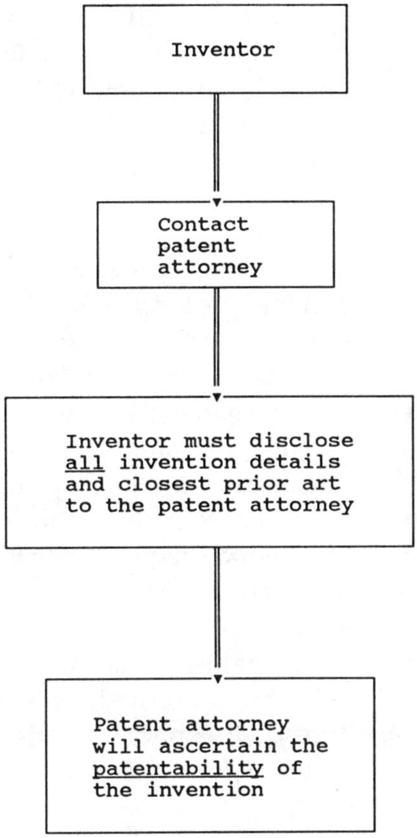

Figure 9.2 The path toward patenting an invention.

other professionals. Professionals in large metropolitan areas with high office expenses can be expected to charge higher fees than comparable professionals in less populated areas.

Having engaged the services of a patent attorney, the inventor should not just sit back and let the patent attorney carry the ball. Rather, the inventor and patent attorney should now begin a cooperative effort toward the goal of obtaining a patent for the invention. The inventor must first give the attorney a complete description of the invention. This description should include all the operative details that would be necessary for a person skilled in the particular art of the invention to make and use the invention without undue experimentation. Figure 9.2 summarizes these preliminary steps. Under United States law, patent applications must disclose the preferred method of practicing a claimed invention; for example, an inventor cannot claim a new process without disclosing the conditions which he or she knows works best. Therefore, the inventor must disclose to the patent attorney the *best mode* contemplated by the inventor, at that time, for practicing the invention. The completed description of the invention provided by the researcher is essential for the patent attorney to prepare a suitable patent application for the invention; this will be discussed in detail in the next chapter.

In addition, in biotechnology there are often materials, such as microorganisms, which require special attention from the researcher and the patent attorney. For example, if a *novel* microorganism is used in the invention, steps must be taken to insure that the microorganism would be available to the public if and when a patent is granted. This can be done by depositing a subculture of the microorganism in a suitable depository (see Appendix B for a list of depositories and organisms accepted by each).

9.1 The Invention Disclosure Format

In assembling the necessary data for presentation to the patent attorney, the inventor will find that using a so-called "Invention Disclosure" format will aid greatly in assembling the data required by the patent attorney. An invention disclosure can be in any of various formats so long as all the necessary information is contained in the disclosure. Two examples of detailed invention disclosures, one for a microbiological invention and one for a genetic engineering invention, are given in Appendix A. The information needed is also summarized in the following list:

1. *Species*: Here the term "species" refers to a specific embodiment (not a taxonomic classification). The inventor needs to describe the specific process, substance, or device which he or she has invented. For example, the inventor may disclose a new fermentation technique which results in higher yields of a product. The technique could be, for example, the addition of a chemical compound to the fermentation broth.
2. *Extensions and Variation*: Next, the inventor delineates other specific extensions or variations of the research work. In order to do this, the inventor should draw on his or her general expertise in the field, as well as the specific research results. For example, the inventor of a new fermentation technique may know of several microorganisms which show increased yields. Also, the technique for increasing the yield may be capable of modification. For example, there may be other chemical compounds which produce the same result.
3. *Genus*: Here the term "genus" refers to a category of related items (not a taxonomic classification). The inventor attempts to find a general category to which all the species of parts 1 and 2 above belong. For example, the researcher may be able to say that the new fermentation technique works with some general class of microorganisms. Or, possibly, there could be a class of chemical compounds which produces the higher yield.
4. *Additional Species*: Once the inventor has formulated the genus, then it should be possible to fill in addi-

tional species which belong to that genus. For example, if the original chemical compound that produced the higher yield was methylamine, and the inventor listed isopropylamine and butylamine as variations, then the inventor might well generalize this to primary amines as the genus. Once the primary amines are established as the genus, other species such as ethylamine and propylamine can be cited as additional specific examples of the genus.

5. *Method*: The inventor must provide a complete description of how to practice the invention. If unique materials are needed to practice the invention, the researcher must disclose a method for producing or otherwise obtaining the materials.
6. *Utility*: The inventor should also describe what he or she perceives to be the use or uses of the invention. For example, the new fermentation technique could be used to produce increased yields of an antibiotic. The utility should be applicable for all members of the genus.

A generic conception by itself is of little value—it must be accompanied by representative species. But a generic conception coupled with representative examples can be invaluable. Therefore, it is very important to include variations and extensions of the initial conception.

The invention disclosure is *not* a *legal* document: it is prepared by scientists, not lawyers. Although the document should be prepared carefully, it should not be construed to be dispositive of *legal* questions. For example, inventorship is a legal question, and the inventors who happen to be named on the invention disclosure may or may not eventually be named as inventors on a patent for that invention. Unfortunately, many such disclosures use the word "inventors" and provide spaces for the researchers to fill in their names. Researchers should recognize that the determination of inventorship of any invention is a legal determination. Thus, a patent attorney should make this finding after all the details of the invention are reviewed.

Therefore, the invention disclosure should merely list the names of the scientists who "submitted" or "prepared" the invention disclosure. Any presumption that such persons are, in fact, the inventors, is in error. Unfortunately, the situation often arises where a scientist has prepared an invention disclosure, and merely filled in the space for "inventor" with his or her name; then he or she is subsequently faulted by associates for omitting their names. This is clearly a misunderstanding of the nature of the invention disclosure document. Avoidance of any problems concerning inventorship can be achieved by simply explaining the purpose of the invention disclosure document to the scientists, deleting the word "inventor" from the document, and substituting a list of the names of the people involved.

One approach to invention disclosure forms involves the use of forms which specifically require information relating to each phase of a project. Examples of these forms are shown in Appendix A. Several comments about these forms will help the inventor understand the need for the information requested. Initially, it is probably clear to most that the forms ask for more information than is necessary to practice or repeat the invention. There are several reasons for asking for this additional data. One reason is that the patent attorney will want to describe the invention in the patent application in a way that would maximize the patent protection for the invention. Merely reciting the precise conditions and materials of the invention would, in most instances, not provide the inventor with good patent protection. The reason for this is that once the patent is granted, a person skilled in the art would be more likely to "work around" the claimed invention without infringing it if the claims are narrow in scope. Therefore, the precise invention materials and/or conditions must be extrapolated to a reasonable degree to cover what is believed to be equivalent materials and/or conditions.

Therefore, much of the information requested in the forms in Appendix A is for the purpose of enabling the patent attorney to broaden the invention disclosure in the patent application. This aspect of patent application drafting is critical for the proper patent protection of most, if not all, biotechnology inventions. It should be stated that the allowed breadth, or scope, of patent claims is also one of the most contested areas of patent law. Often the patent attorney and the patent examiner will not agree on the permissible scope of claims. Many of these differences are exposed in appellate court decisions. A patent attorney with experience in prosecuting biotechnological patent applications, and who is knowledgeable about relevant court decisions, is best equipped to work with inventors in an effort to properly broaden the scope of the claimed biotechnological invention.

Other information requested in the invention disclosure is important from the standpoint of having a record of the critical dates in the research leading to the invention. This record will be very useful when, and if, an interference proceeding is conducted at a later date. As previously discussed in Chapter 3, an interference proceeding is conducted by the PTO when two or more pending patent applications (or a patent application and an issued patent) claim essentially the same invention. The purpose of the interference proceeding is to determine the identity of the *first* inventor of the invention in issue. This determination of the first inventor is done on the basis of the research records.

By listing dates and notebook identifications in the forms of the invention disclosure, as in Appendix A, the inventor has provided the patent attorney with what could turn out to be very crucial information if an interference is subsequently declared. The patent attorney could, if necessary, ferret out the necessary information at the time an interference is declared. However, this would be very difficult and time consuming, especially because most interference proceedings are not initiated until several years after a patent application has been filed in the PTO. This long hiatus between the filing of a patent application and the declaration of interference proceedings is due in great part to the complexity of many patent applications. This is particularly true with respect to biotechnology inventions. Also, the enormous number of applications filed causes a backlog at the Patent Office. Therefore, the time elapsing between the filing of an application and the declaration of interference proceedings can be three or four years or more.

There is also the possibility of an interference between an issued patent and a pending patent application. Such an occurrence, again, can occur several years after the filing of a patent application. Therefore, if the inventor records events and dates in an invention disclosure when all these data are still relatively fresh in mind and readily available, the inventor has taken a giant step toward fully protecting the invention.

From the above discussion it should be readily apparent that invention records should be kept in a file in a safe place for a time sufficient to allow all chances of an interference proceeding to expire. Records which should be kept include notebooks, reports, and pertinent memos and letters. Generally, records of an invention should be safeguarded for at least three years after a patent for the particular invention issues.

Two of the forms of the invention disclosure in Appendix A (numbers 15 and 16) are used to inform the patent attorney of any relevant prior publications and possible prior use of the particular invention. This information will be studied by the patent attorney to see if there is anything which might tend to cast a cloud over the patent protection sought. For example, if the invention was used commercially for more than one year, then, under most circumstances, there would be a *bar* to obtaining a patent. However, each situation would have to be explored in detail by the patent attorney because there are a variety of exceptions to the bar. This is also true with regard to prior publications, as was discussed in Chapter 5. For example, a prior publication of an invention may not be detrimental to patenting the invention if the publication was not an enabling disclosure. There are also other exceptions, and all should be addressed by the patent attorney.

The patent attorney needs to know about prior publications and uses not only to evaluate the potential for receiving a patent, but also because the patent attorney *must* disclose relevant prior publications and uses to the PTO when the patent application is filed. Any avoidance of this responsibility can lead to serious consequences for both the inventor and the patent attorney. Failure to disclose prior publications and uses in the patent application can invalidate the patent in question and the attorney can be disbarred. Thus, it is very important for the researcher to be as thorough as possible in completing the invention disclosure.

The forms shown in Appendix A relating to genetic engineering inventions are used to elicit essentially the same kind of information as requested in the forms for a microbiological invention. These disclosure forms will give the patent attorney sufficient initial data to appraise the patentability of the invention, as well as the potential for broad claims. It should be known, however, that broadening the scope of a *genetic engineering* invention is not a simple matter. The difficulty encountered in this regard is due largely to the perceived relative unpredictability in the genetic engineering art. When a particular art is deemed to be unpredictable, the Patent Office will limit the scope of claims so that only results which have been empirically obtained will receive patent protection. The applicant is not allowed to generalize his or her results in order to make broad claims. As confidence grows in the predictability of this art, then the scope of broadening should be more extensive. These developments can be expected as further work is done and there is a better understanding of the basic events in genetic engineering. However, at the present time, the facts arguing for predictability are not believed to be strong in many areas of this art.

With regard to the predictability of the genetic engineering art, a frustrating scenario frequently occurs during the process of prosecuting applications for genetic engineering inventions. Frequently, an applicant will be required to limit his or her claims so that they cover only results which have been empirically obtained. As mentioned above, the rationale for limiting the scope of claims in this manner is that the art is too unpredictable to allow extrapolation. Surprisingly, however, many of these same genetic engineering inventions are rejected by the Patent Office on the grounds that they would have been obvious to one skilled in the art. It is not at all clear how so many inventions can be obvious in such an unpredictable art. Hopefully, as the law with respect to genetic engineering inventions begins to develop more fully, the use of this contradictory reasoning will diminish.

9.2 Patentability and Commercial Utility

The nature of an invention disclosure need not vary extensively whether the disclosure is used by a corporate researcher or by an academic researcher. In a corporate system, the invention disclosure is generally examined by the patent law department of the corporation. This department, staffed by patent attorneys, can readily evaluate the patentability of the invention, but it is best if the patent law department is not also responsible for assessing the commercial utility of the invention.

Patent attorneys are trained to assess patentability, not commercial utility. This is not to say that the patent law department should not be consulted on any commercial evaluation. However, being consulted is completely different from having the responsibility for the ultimate decision on whether to file a patent application. Rather, the assessment of commercial potential should be done by persons skilled in the science of the invention and by marketing experts. It is also usually advisable for the inventor to be a member of the group making this determination.

In the case of biotechnology inventions, and especially for genetic engineering inventions there are almost no market data available for assessing commercial worth of an invention. Therefore, it is more correct to speak about *potential* commercial utility, rather than actual commercial utility. Thus, use of only standard market assessment techniques, as used, for example, in marketing a new cereal or a bar of soap, would probably result in an inaccurate prediction of market potential. We all know that many of the products of genetic engineering research will, in fact, create new markets. These potential markets can only be assessed by the combined efforts of scientists knowledgeable in the art and by marketing experts.

For an invention in an academic environment, the invention disclosure is generally first examined by a patent committee composed of scientists. The scientists try to make an initial determination of the commercial potential of the invention. Whether such a group can be reasonably expected to make an accurate determination will depend largely on the marketing experience of the members of the group. If the patent committee feels that the invention has potential commercial utility, the invention disclosure will pass to the patent law department for the next step of evaluation. At this time, a patent attorney will use the invention disclosure to make a determination of *patentability*. If the decision is that the invention is patentable, then a patent application will usually be filed. This procedure is satisfactory only if the members of the initial patent committee have sufficient marketing knowledge.

10

Depositing a Novel Microorganism

The filing of a patent application for a biotechnology invention involving the use of a novel microbe necessitates making the microbe available to certain parties. During the time that the patent application is pending in the United States PTO, the microbe must be available to the Patent Office. Generally, the Patent Office has no need to examine these microorganisms. However, during certain PTO proceedings, such as interference proceedings, the Patent Office has the power to make microbes available to contesting parties. Accordingly, the Patent Office must be in the position to so act.

Also, upon the granting of a U.S. patent, the microbe *must* be available to the public.

In order to make a novel microorganism readily available, there has evolved in the biotechnology art a procedure of depositing the novel microbe in a suitable culture depository. In the United States the two depositories that are commonly used are the American Type Culture Collection (ATCC) in

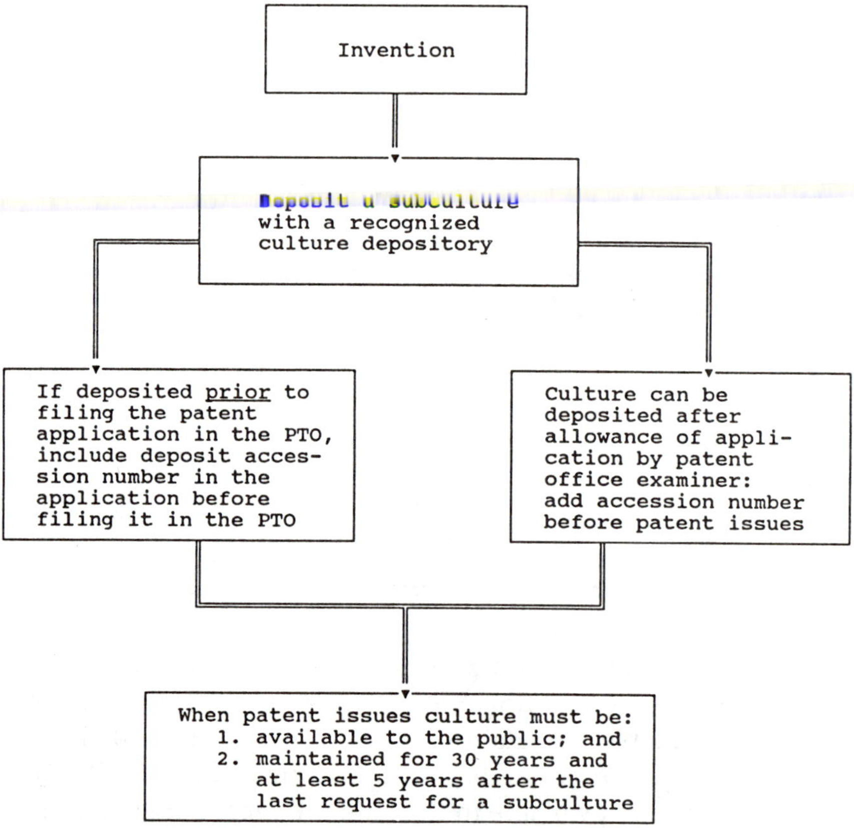

Figure 10.1 Two paths for depositing a culture in order to obtain a patent for a novel microbe.

Rockville, Maryland, and the Agricultural Research Culture Collection of the USDA (known as the *NRRL*) in Peoria, Illinois. When a deposit is made at one of these depositories, an accession number is assigned to the culture. By referring to this accession number, it is possible for interested parties to receive subcultures of the novel microorganism. By referring to the accession number in the text of a patent document, a patentee effectively makes that novel microorganism available to the public. Although these deposits are generally made prior to the filing of a patent application, under United States patent law it is possible to deposit the culture and add the accession number *after* filing but *before* the patent issues. Figure 10.1 depicts the sequence of events with regard to the deposit of a novel microorganism.

For making a deposit with a culture depository, certain procedures have been developed that are accepted by the U.S. PTO and by patent offices throughout the world. Figure 10.2 is an example of a culture deposit form letter that can be used for depositing a novel microbe with the Agricultural Research Culture Collection (NRRL). The *Budapest Treaty* referred to in the deposit letter was the result of a meeting held in Budapest, Hungary in 1977, and has been signed by most of the industrially important countries of the world. This treaty provides that a culture deposit in an acceptable culture depository will be recognized by all member countries that require culture deposits in conjunction with patent application filings. This, then, obviates the necessity of having to make deposits in every country where a patent application is filed. The Republic of Korea is the latest country to accede to the Budapest Treaty. Its membership became effective March 28, 1988.

The letter shown in Figure 10.2 (pages 96 and 97) should be used to accompany the Budapest Treaty form, shown in Figure 10.3 (pages 98 and 99), entitled "Microorganism Deposit Under the Budapest Treaty on the International Recognition of the Deposit of Microorganisms for the Purposes of Patent Procedure." It is best for an inventor to deposit a novel microbe using this treaty procedure, because the treaty is accepted both in the United States and in all other major patent systems throughout the world. Further details for depositing a microbe can be obtained from the patent attorney who handles the filing of the patent application.

Mr. A.J. Lyons, Curator
ARS Patent Collection
Culture Collection Research-
Fermentation Laboratory
Northern Regional Research Center
Peoria, Illinois 61604

Dear Mr. Lyons:

Under separate cover, we are sending to you the following (check one):

[ ] two agar slants
[ ] thirty lyophilized preparations
[ ] two lyophilized preparations

of the following viable organism:

[organism being deposited]

This organism is being deposited under the Budapest Treaty on the International Recognition of the Deposit of Microorganisms for the Purposes of Patent Procedure. It is to be added to your permanent collection, i.e., to be maintained for at least thirty (30) years. Should the culture be lost, replacement thereof will be made with additional agar slant transfers or lyophilized preparations. However, it should be known that the culture was viable when it was sent to you.

Please withhold distribution of the culture, except under the conditions hereafter specified. Access to this culture is hereby granted to:

1. the United States Commissioner of Patents and Trademarks;
2. such individuals as the United States Commissioner of Patents and Trademarks shall designate;
3. such individuals whom we designate to you in writing;
4. the public upon the issuance of a United States Patent to [name of applicant or assignee], which patent identifies this culture by the NRRL number assigned to it;
5. the public upon the publication of any patent application which application identifies this culture by the NRRL number assigned to it.

Figure 10.2 An example of a form letter used for making a culture deposit.

Upon receipt of the microbe by the culture depository, the curator of the depository will send the depositor a receipt and a viability statement, to be completed by the depositor and returned to the curator. The receipt form and viability statement are shown in Figures 10.4 (page 100) and 10.5 (page 101).

All restrictions on the availability to the public of the above culture deposit will be irrevocably removed upon the granting of a United States patent disclosing the deposit.

Please sign the accompanying form on the designated line, return the original to us, and keep a copy for your files.

Enclosed is a check made out to the Agricultural Research Service, USDA for $500.00 (five hundred dollars). Also enclosed are the executed Budapest Treaty forms.

Very truly yours,

[name]

ACKNOWLEDGEMENT OF RECEIPT AND
DESIGNATION OF DEPOSIT NUMBER

[ ] Two agar slants
[ ] thirty lyophilized preparations
[ ] two lyophilized preparations of

[deposited culture],

and the $500.00 check, have been received, and we agree to the above-mentioned conditions.

The culture slants were received on: ____________________.

The NRRL number is: ____________________.

Signed: ____
A.J. Lyons   Date

Please return the signed original; keep a copy for your files.

The Budapest Treaty deposit form for the American Type Culture Collection (ATCC) is reproduced in Figure 10.6 (pages 102 and 103). Any questions concerning the deposition of cultures in the ATCC depository in conjunction with patent applications can be directed to the curator, at the address and phone number given on the form.

AGRICULTURAL RESEARCH CULTURE COLLECTION (NRRL)
(International Depositary Authority)

Fermentation Laboratory, Northern Regional Research Center
Science and Education Administration, U.S. Department of Agriculture
1815 North University Street, Peoria, Illinois 61604 U.S.A.

ADDRESS SHIPMENTS AND FORMS TO: Mr. A. J. Lyons at above address

MICROORGANISM DEPOSIT UNDER THE BUDAPEST TREATY ON THE INTERNATIONAL RECOGNITION OF THE DEPOSIT OF MICROORGANISMS FOR THE PURPOSES OF PATENT PROCEDURE (Budapest Treaty)

THIS BOX FOR NRRL USE ONLY

Date and Nature of Material Received: ____________________

Date Deposit Accessioned: ____________________

as NRRL No. ____________________

*(Contingent on depositor's verifying acceptability of Agricultural Research Culture Collection (NRRL)'s processed material as equivalent to original material deposited)

DEPOSIT STATEMENT

THIS FORM MUST BE COMPLETED IN ENGLISH

[Our present policies are stated in Industrial Property No. 1, pp. 22-24 (1981)]

1. Name of Microorganism and Depositor's Strain Designation (acronym, sigla, abbreviation + number, symbols)

____

This microorganism is: a bacterium ☐, an Actinomycetales ☐, a mold ☐, a yeast ☐, an alga ☐, a strain containing recombinant DNA molecules ☐, a strain containing its own naturally occurring plasmid(s) ☐, a strain containing inserted naturally occurring plasmid(s) from another host ☐, a strain containing inserted constructed plasmid(s) ☐, a strain containing a virus of any kind ☐.

2. Is this strain being deposited under the Budapest Treaty? ____________________

3. Name and Address of Depositor:* ____________________

____

____

____

____

*Viability test reports and other correspondence will be sent to above party. Our processed material from the original deposit will be sent to above party for checking. If there is no response within 3 months from date of shipment, our processed material will be considered equivalent to the original deposit in terms of viability and performance attributed to the strain.

4. Indicate the properties of the microorganism which are or may be dangerous to health or the environment, or indicate that the depositor is not aware of such properties.

____

____

5. Is this strain zoopathogenic? ______ phytopathogenic? ________

6. Recommended Conditions for Optimal Cultivation of the Microorganism and for Testing Its Viability:* ______________________________

____

____

____

*Progeny of strains will be preserved at the Agricultural Research Culture Collection (NRRL) as lyophilized preparations or, in some cases, as agar slant cultures overlayed with mineral oil. All materials will be stored at 3-5° C.

7.* Lowest permissible NIH** (U.S.A.) Physical Containment level for processing and viability testing (P1, P2, P3, P4):
Lowest permissible CDC*** (U.S.A.) Biosafety level for processing and viability testing (1, 2, 3, 4): ________

* The Agricultural Research Culture Collection (NRRL) can, at present, process progeny of strains <u>only at Physical Containment level P1 or Biosafety Level 1 or less</u>.

** U.S. Department of Health and Human Services, National Institutes of Health, Bethesda, Maryland 20205, U.S.A., November 1980. Guidelines for Research Involving Recombinant DNA Molecules.

*** U.S. Department of Health and Human Services, Public Health Service, Centers for Disease Control, Office of Biosafety, Atlanta, Georgia 30333, U.S.A. 1980. Proposed Biosafety Guidelines for Microbiological and Biomedical Laboratories.

8. Are you willing to waive the right to be informed of all requests for progeny of this strain? (This is allowed under the Budapest Treaty but will require additional correspondence.) ____________________________

9. I understand and agree that the deposit may not be withdrawn by me or any representative of my organization for the period specified in Rule 9.1 of the Budapest Treaty (at least 30 years after the date of accessioning).

Date __________________ Signature of Depositor ______________________

__________________ (on behalf of) ______________________ (Typed name of depositor)

Figure 10.3 A Budapest Treaty form for depositing a culture with the NRRL.

BP/A/II/12
page 14

BUDAPEST TREATY ON THE INTERNATIONAL
RECOGNITION OF THE DEPOSIT OF MICROORGANISMS
FOR THE PURPOSES OF PATENT PROCEDURE

INTERNATIONAL FORM

TO

NAME AND ADDRESS
OF DEPOSITOR

RECEIPT IN THE CASE OF AN ORIGINAL DEPOSIT
issued pursuant to Rule 7.1 by the
INTERNATIONAL DEPOSITARY AUTHORITY
identified at the bottom of this page

I. IDENTIFICATION OF THE MICROORGANISM	
Identification reference given by the DEPOSITOR:	Accession number given by the INTERNATIONAL DEPOSITARY AUTHORITY:

II. SCIENTIFIC DESCRIPTION AND/OR PROPOSED TAXONOMIC DESIGNATION
The microorganism identified under I above was accompanied by: ☐ a scientific description ☐ a proposed taxonomic designation (Mark with a cross where applicable)

III. RECEIPT AND ACCEPTANCE
This International Depositary Authority accepts the microorganism identified under I above, which was received by it on (date of the original deposit)[1]

IV. INTERNATIONAL DEPOSITARY AUTHORITY	
Name: Agricultural Research Culture Collection (NRRL) International Depositary Authority Address: 1815 N. University Street Peoria, Illinois 61604 U.S.A.	Signature(s) of person(s) having the power to represent the International Depositary Authority or of authorized official(s): Date:

[1] Where Rule 6.4(d) applies, such date is the date on which the status of international depositary authority was acquired; where a deposit made outside the Budapest Treaty after the acquisition of the status of international depositary authority is converted into a deposit under the Budapest Treaty, such date is the date on which the microorganism was received by the international depositary authority.

Form BP/4 (sole page)

Figure 10.4 A Budapest Treaty receipt issued by the NRRL.

BUDAPEST TREATY ON THE INTERNATIONAL
RECOGNITION OF THE DEPOSIT OF MICROORGANISMS
FOR THE PURPOSES OF PATENT PROCEDURE

INTERNATIONAL FORM

TO

VIABILITY STATEMENT
issued pursuant to Rule 10.2 by the
INTERNATIONAL DEPOSITARY AUTHORITY
identified on the following page

NAME AND ADDRESS OF THE PARTY
TO WHOM THE VIABILITY STATEMENT
IS ISSUED

I. DEPOSITOR	II. IDENTIFICATION OF THE MICROORGANISM
Name: Address:	Depositor's taxonomic designation and accession number given by the INTERNATIONAL DEPOSITARY AUTHORITY: Date of: ____________ [1] ☐ [2]Original Deposit ☐ [2]New Deposit ☐ [2]Transfer

III. VIABILITY STATEMENT AND DEPOSITOR'S EQUIVALENCY DECLARATION

Deposit was found: ☐ [2]Viable ☐ [2]Nonviable on ____________ (Date)
International Depositary Authority's preparation was found viable on ____________ (Date)
Depositor determined the International Depositary Authority's preparation was
☐ [2]Equivalent ☐ [2,3]Not Equivalent to deposit on ____________ (Date)

Signature of Depositor ____________

[1] Indicate the date of the original deposit or when a new deposit or a transfer has been made.

[2] Mark with a cross the applicable box.

[3] If found nonequivalent, depositor must supply 30 lyophilized preparations made to NRRL size specifications.

IV. CONDITIONS UNDER WHICH THE VIABILITY TEST WAS PERFORMED[4]

V. INTERNATIONAL DEPOSITARY AUTHORITY

Name: Agricultural Research Culture Collection (NRRL) International Depositary Authority Address: 1815 N. University Street Peoria, Illinois 61604 U.S.A.	Signature(s) of person(s) having the power to represent the International Depositary Authority or of authorized official(s): Date:

[4]Fill in if the information has been requested and if the results of the test were negative.

Figure 10.5 A Budapest Treaty viability statement issued by the NRRL.

Budapest Treaty Deposits

American Type Culture Collection

12301 Parklawn Drive • Rockville, MD 20852 USA • Telephone: (301)881-2600 Telex: 898-055 ATCCNORTH

TO DEPOSIT OR TO CONVERT A DEPOSIT TO MEET THE REQUIREMENTS OF BUDAPEST TREATY ON THE INTERNATIONAL RECOGNITION OF THE DEPOSIT OF MICROORGANISMS FOR THE PURPOSES OF PATENT PROCEDURE

*1. Name of deposit (microorganism, cell, seed, plasmid, etc.) ____________

2. Strain designation given by the depositor (number, symbols, etc.) ____________

3. Is this an original deposit under the Budapest Treaty? ____________

4. Is this a request for a conversion of a deposit already at the ATCC to meet the requirements of the Budapest Treaty? (If so, indicate ATCC designation.) ____________

5. Is this deposit a mixture of microorganisms or cells? ____________

6. Details and conditions necessary for the cultivation of the strain, for its storage and for testing its viability and also, where a mixture of microorganisms is deposited, descriptions of the components of the mixture and at least one of the methods permitting the checking of their presence. ____________

7. An indication of the properties of the strain which are or may be dangerous to health or the environment, or an indication that the depositor is not aware of such properties. ____________

*8. It is recommended that sufficient description be provided to allow the ATCC to confirm that the strain deposited generally conforms to that which the depositor states is being deposited (*i.e.*, Gram negative rod). ____________

   a. For cell culture deposits please complete. Is the cell being cultured in the presence of antibiotics (if so list the antibiotics) ____________

   b. For hybridoma deposits please complete. What is the isotype of antibody produced? ____________

*9. Is this strain zoopathogenic? ____________ phytopathogenic? ____________

10. Does this strain contain plasmids relevant to the patent process? ____________
If so, what physical containment level is required for experiments as described in the National Institutes of Health Guidelines involving Recombinant DNA Molecules (*i.e.*, P1, P2, P3 and P4 facility)? ____________

*The answers to these questions are recommended but not required.

FOR ATCC USE ONLY

ATCC DESIGNATION ____________

DATE CULTURE RECEIVED ____________

DATE VIABILITY TEST COMPLETED ____________

Form BP/1 (Page 1 of 2)

Figure 10.6 A Budapest Treaty form for depositing a culture with the ATCC.

*11. Isolated from? ____

____

12. In addition to those entitled to a sample under the Budapest Treaty, do you wish the strain made available to (Those entitled to a sample under the European Patent Convention and countries signatory to the Budapest Treaty will receive samples.):

a. Anyone who requests a culture (no restrictions on distribution from date of deposit or conversion to Budapest)?

____

b. Other countries which are not signatory to the Budapest Treaty? Please state which countries: ______________

____

After a U.S. Patent issues, the ATCC makes the culture available to anyone who requests it.

13. Do you wish the ATCC to inform you of all requests for this strain? (This is allowed under the Treaty, but if you waive the right, the fee is reduced.) ______________________________

14. Would you like to be notified via telex or telephone of the ATCC number assigned to your strain (Fee: $10.)? ________

Telex No. ______________________ Telephone No. ______________________

15. Viability testing certificate and contracts should be directed to:

____

____

____

16. Check in payment of service must accompany the deposit unless prior arrangements are made and approved. If arrangements have been made to bill you for services, an invoice should be sent to:

____

____

17. Name and address of attorney (to whom information will be made available if you complete): ______________

____

18. Additional comments: ____

____

I understand and agree that the deposit may not be withdrawn by me for the period specified in Rule 9.1 of the Budapest Treaty (at least 30 years after the date of deposit), and that if a strain should die or be destroyed during the life of the patent, or the period of time so specified, it is my responsibility to replace it with a living culture of the same organism or cell. In the cases of viruses, cell cultures, plasmids and seeds, it is my responsibility to supply a sufficient quantity for distribution for the period of time specified above.

DATE ______________ SIGNATURE OF DEPOSITOR ______________________

____

On behalf of (Name of company or institution if signed on their behalf) **(must be completed)** (Typed name of Depositor)

Address: ______________________

____

____

THIS FORM MUST BE COMPLETED IN ENGLISH

ADDRESS SHIPMENTS AND FORM TO THE ATTENTION OF:

Mrs. Bobbie A. Brandon
American Type Culture Collection
12301 Parklawn Drive
Rockville, Maryland 20852 USA

*The answers to these questions are recommended but not required.

Form BP/1 (Page 2 of 2) 3/87

Although the NRRL culture depository in Peoria, Illinois, is the example used in the above discussion and Figures 10.2 through 10.5, the inventor may use any of the accepted culture depositories located in the United States and throughout the world. No matter which depository is used, the advantage of depositing under the Budapest Treaty is that only one deposit is needed to fulfill the patent requirements of a number of countries. Prior to the institution of this single culture deposit system, a biotechnology inventor was required to deposit a microbe in a culture depository in virtually every major country where a patent application for the invention was filed. This former system was clearly much more expensive and time-consuming. Appendix B lists approved culture depositories, alphabetized by country, and the types of deposits which each one will accept.

11

The Patenting Path in the United States

Once a patent application is filed with the United States Patent and Trademark Office (PTO), a series of events ensues that can lead to the granting of a patent. This series of events is collectively known as the *prosecution* of the application. It is desirable for the inventor to have some knowledge of these events because this knowledge may help alleviate some of the anxiety which an inventor may have while the patent application is pending in the PTO. In addition, this knowledge will help the inventor to aid the patent attorney during the prosecution of the patent application before the Patent Office. Figure 11.1 diagrams the steps leading to a patent in the United States. These steps will be discussed in detail in this chapter. Figure 11.1 also includes the steps that are followed in order to continue patent prosecution after an application is given a "final rejection" by the Patent Office.

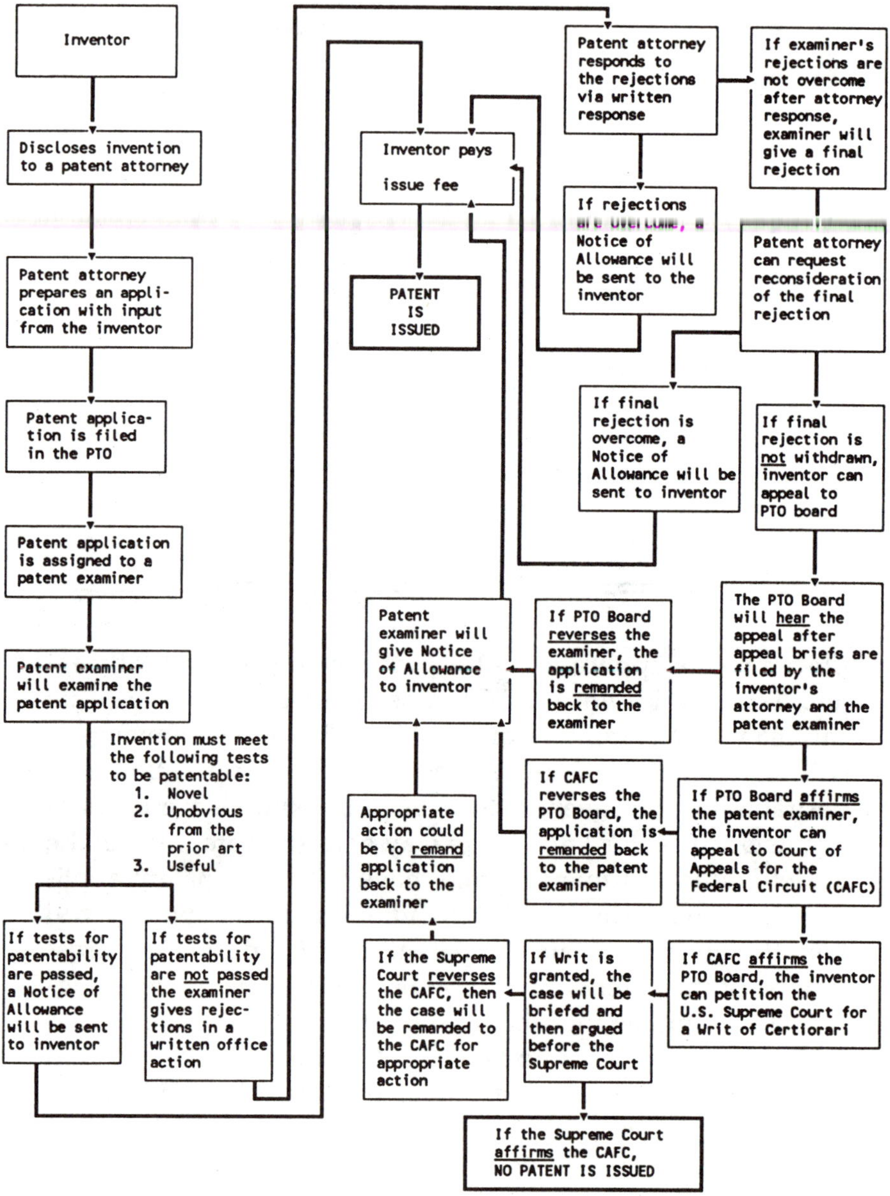

Figure 11.1 The complete patenting path in the United States.

11.0 The Patent Examiner

Once a patent application has been prepared, the first step in initiating the examination of that application is to file the application with the Patent Office. The application which is sent to the Patent Office must be accompanied by certain formal documents and the proper filing fee. The filing fee consists of a basic fee plus an additional charge according to the number and types of claims contained in the application. Currently, the basic filing fee is $340.00. However, if the applicant qualifies as a *small entity* or a *non-profit organization,* then the basic filing fee is only $170.00. In order to qualify as a small entity, a business cannot have more than 500 employees.

After the PTO receives a patent application which is accompanied by the appropriate fee, the application is given a serial number and a filing date. The application is then routed to the particular examining group in the PTO which examines the type of invention to which the patent application is directed. There are over one hundred such examining groups in the Patent Office. After arriving in the group in charge of examining the particular type of invention, the director of the examining group will assign the application to one of the art units, that is, the examining sub-groups, under the group's domain. In the next step, the supervisory patent examiner of the particular art unit will then assign the patent application to a specific examiner. A patent examiner is a person who is technically trained in the subject matter pertaining to patent applications assigned to him or her. The examiner is also trained in patent law, but may or may not be a lawyer. The examiner's primary function is to make the determination as to whether an invention is patentable under the patent statutes.

The ultimate assignment of a patent application to a patent examiner is designed to place the patent application in the hands of a patent examiner who is knowledgeable in the art to which the invention pertains. For example, a patent application relating to a biotechnology invention will be assigned to an examiner knowledgeable in biotechnology. Preferably, because the biotechnology art is rather broad, the examiner receiving a patent application for examination will be trained in the particular art to which the invention pertains. Thus, for example, a patent application for a recombinant DNA invention will be assigned to an examiner able to understand such an invention.

When a patent examiner receives a new patent application from the supervisory patent examiner, the application will be placed at the bottom of the pile of other applications awaiting examination. When applications concerning molecular biology inventions first began to be filed in the PTO, examiners working on these applications soon had a large backlog of patent applications awaiting their initial examination. Because of this large backlog, a patent application for a genetic engineering invention would probably not be examined for at least one or two years after the patent application was filed in the PTO. This length of time is not unusual when a large number of applications are filed in the PTO in a short period of time. The large number of biotechnology patent applications filed results from the high research activity in this technology. Although the Patent Office has made tremendous efforts to get caught up with the backlog of biotechnological inventions, an applicant can still expect to wait at least one year from the date of filing for the initial examination of the application.

When a particular patent application is finally up for examination, the patent examiner will first review the application in order to make sure that certain formal requirements have been met. For example, all patent applications must have a specification and one or more claims. The specification generally consists of a title, an abstract, a background of the invention, a brief summary of the invention, a brief description of the drawings, the drawings themselves, and a detailed description of the invention. For a complex invention, the application may be very long: 50 pages or more is not unusual. Even for relatively simple inventions, the application will generally be at least five to ten pages. Not surprisingly, the professional fees associated with the preparation of a patent application will vary depending upon the complexity of the subject matter. Accordingly, the cost of preparing a patent application could range from $2,000.00 for a relatively simple mechanical invention up to $8,000.00 and more for a more complex invention.

Once the examiner is satisfied that the requirements with respect to the format of the application have been satisfied, the next step is to evaluate the invention against the requirements of the Patent Act for a patentable invention. As stated earlier, these statutory requirements are that the claimed invention be (1) novel, (2) unobvious, and (3) useful. Although these requirements may sound straightforward, there are literally hundreds of court decisions on a variety of inventions that set out the legal metes and bounds of these statutory requirements. Clearly, a patent examiner is asked to do a very complicated job, and in a relatively short length of time. For each patent application the examiner must:

1. Read and understand the invention.
2. Search for and read the prior art, that is, patents and publications worldwide.
3. Compare the closest prior art with the invention claimed in the patent application.
4. Decide whether to allow claims or reject claims under one or more of the sections of the patent statute.
5. Communicate examination results via a written document called an *office action,* to the patent attorney who filed the application.

Depending upon the nature of the examiner's response, the office action may range in length from as few as one page to as many as fifteen pages or more. The steps leading to this point are summarized on the left side of Figure 11.1 (see page 106).

Obviously, the patent system would not exist without the dedicated professional work of the patent examiners. Much of their work has to be a labor of love because their work is demanding in terms of hours worked and scientific knowledge, and their financial rewards are good but not great.

11.1 The Office Action

When an examiner has finished evaluating a patent application, he or she may decide that some or all of the claims are patentable. In this case, the examiner is said to "allow" the patentable claims, and a Notice of Allowance will be sent to the patent attorney. Then the inventor only needs to pay the issue fee (see page 112) and the patent will be isssued about three months later.

If, on the other hand, the examiner finds that some, or all, of the claims are not patentable, he or she will reject these claims. When claims are rejected, the examiner will communicate this fact to the patent attorney in an office action. The office action contains the specific rejections as well as the reasons for these rejections. The office action may also contain other comments and/or objections that the examiner may have to the content or format of the patent application. It is very unusual for all of the claims in a patent application to be allowable after the initial evaluation by the patent examiner. Therefore, an inventor should never be surprised when the initial office action contains one or more rejections.

11.2 Responding to the Office Action

After the patent attorney has received the examiner's office action, he or she will generally have three months in which to file a *timely response*. A response can be filed up to three months later by the payment of a fee at the time the response is filed. If the response is not filed within six months of the date of the office action, the application is deemed to have been abandoned. It is sometimes possible to revive an abandoned application. However, in order to do this, it is generally necessary to show that the failure to respond within the six-month period was unavoidable.

When the patent attorney receives the office action, the inventor will once again become involved in the patenting process. The patent attorney will transmit the office action to the inventor, seeking comments concerning the rejections. The inventor will be asked by the patent attorney to comment on any scientific matters raised in the office action. For example, the inventor may be asked to review a publication cited by the patent examiner and to state whether or not the patent examiner is correct in the interpretation of the publication. Even though patent examiners are presumed to be experts in the art to which the invention in issue pertains, it is not realistic to expect them to be infallible in their interpretation of a publication. Just as two scientists can disagree on scientific content and results, an inventor can disagree also with a patent examiner's scientific analysis.

The patent attorney will work with the inventor to prepare a response to the office action. The patent attorney must respond to each of the examiner's rejections in a manner which is reasonably calculated to bring about an allowance of the claims. The attorney's response may include suitable amendments to overcome rejections. Also, appropriate legal arguments will be advanced in an effort to demonstrate the patentability of the rejected claims. The legal arguments will be formulated using scientific input from the inventor.

11.3 Reviewing the Applicant's Response: The Second Office Action

The patent examiner will review the applicant's response to determine whether all of the rejections of the initial office action have been overcome. Because patent examiners will handle anywhere from about fifty to about two hundred applications at any one time, an applicant should not expect the patent examiner to review the applicant's response until about three months to a year after the response is received in the PTO.

If the patent examiner finds that all the rejections of the initial office action have been overcome, he or she will send a Notice of Allowance to the applicant, indicating that when the issue fee is paid, a patent will issue. The issue fee for a small or nonprofit entity is $280.00 and that for a large for-profit entity is twice that amount. The applicant will then have three months to pay the issue fee. Upon receipt of the issue fee, the PTO will begin processing the application toward the granting of a patent. In most cases, a patent will *issue* about three months after the issue fee is received by the PTO. At this time the PTO will send the attorney notification of the patent number and issue date. A few days after the issue date the attorney can expect to receive a certified copy of the patent.

If the patent examiner finds that one or more of the rejections contained in the initial office action has *not* been overcome by the response, the patent examiner will send the attorney a second office action. This second office action, which is normally termed a "final rejection," can either be responded to by the applicant's attorney, as above, or the applicant can *appeal* the patent examiner's final rejection to the PTO Board of Appeals and Interferences (PTO Board), as shown on the right side of Figure 11.1 (see page 106).

However, if the second office action is *not* termed a final rejection, the applicant can respond as was done after receiving the initial office action. Whether the second office action is a final rejection is at the examiner's discretion. Appeal to the PTO Board is possible only after the patent examiner has given a final rejection.

11.4 Digest of Events After Prosecution Before the Patent Examiner

The above examination procedures before the patent examiner are referred to as "prosecution" before the patent examiner. Once the patent examiner has completed the examination phase, either an allowance or final rejection is given. Recourse from a final rejection is initially to the PTO Board of Appeals.

Proceedings before the PTO Board involve the preparation of appellate briefs by both the attorney for the patent applicant and the patent examiner. These briefs are structured legal documents which place the pertinent facts before the PTO Board for its evaluation and subsequent decision. As a part of this appeal procedure, the applicant's patent attorney can present an oral argument before the PTO Board. The patent examiner can also present an oral argument, but this is not usually done.

The Appellate Board for any particular case consists of three Examiners-in-Chief. Due to the large backlog of cases, the Board will generally decide a case from about one to three years after an appeal is filed. During this time the appellate briefs are written, oral arguments may be given, and the Board eventually reaches its decision. The Board's decision is generally handed down one to three months after the oral hearing.

The procedure before the PTO Board is not open to the public. Thus, if an applicant wishes to maintain the subject matter as a trade secret after a PTO Board decision, this can be done by merely abandoning the patent application. Of course, the other requirements for a trade secret must still be in place. These have been extensively discussed in previous chapters.

If the PTO Board *reverses* the patent examiner's rejections, the patent application is sent back, or remanded, to the patent examiner. The patent examiner will then take procedural steps toward allowing the claims in the application. Subsequently, the applicant can pay the issue fee and, thus, start the procedure for the issuance of a patent.

Affirmation of the patent examiner's rejections by the PTO Board still leaves the applicant with an option to appeal further to the Court of Appeals for the Federal Circuit (CAFC) as shown at the bottom of Figure 11.1. However, appeal to the CAFC opens the subject matter of the application to public view because the public can obtain briefs filed before the CAFC and also attend the oral hearing before the court.

As was done for the PTO Board, the applicant prepares an appellate brief for the CAFC. The solicitor of the PTO represents the patent examiner and the PTO. The solicitor, who is trained as a patent attorney, files an appropriate brief and argues the appeal opposite the applicant's attorney at the oral hearing.

The CAFC uses panels consisting of three judges to handle appeals. Generally, a decision from the CAFC can be expected within about one year after the appeal is filed with the court. If the CAFC reverses the PTO Board, the application is remanded to the patent examiner for disposition in accordance with the court's decision.

Affirmation by the CAFC of the PTO Board decision leaves only the Supreme Court of the United States of America as the next step in the appellate process. The Supreme Court chooses to hear very few patent cases. Thus, for most practical purposes, an adverse decision by the CAFC spells the end of the patenting path.

The above is but a sketch of the complicated appeal process in place in the United States. Certainly, most patent applicants will not go beyond the PTO Board. Some applicants who feel they have important inventions or significant issues of law, will go as far as the CAFC. Most significantly, the appellate process is not required for a large number of inventions for which patents are granted every year in the United States.

Appendix A

Two Invention Disclosure Formats

1. For a Microbiological Invention

One effective way to disclose a microbiological invention is by filling out *forms* for each of the individual parts of the information needed. A set of such forms is as follows:

Form 1. Initial Isolation and Screening of a Biologically Pure Culture

- **1A.** Date first isolated as a biologically pure culture.
- **1B.** Isolate number.
- **1C.** Notebook reference.
- **1D.** Person(s) who did the work.
- **1E.** Person(s) who witnessed and understood the work being done.

Form 2. Characterization of the Biologically Pure Culture

2A. Give a complete art-accepted taxonomic description of the biologically pure culture. If the taxonomy has not been completed, then give the data already obtained. The remainder should be submitted to the patent attorney as soon as it is obtained.

Form 3. Fermentation Development

3A. The development of the fermentation and scaling-up was (check either no. 1 or no. 2):

[ ] **1.** Conducted using standard media and techniques

[ ] **2.** Conducted using special media and/or techniques

If no. 2 is checked, give

a. Fermentation problems incurred which required the use of other than standard media and/or conditions

b. Media and/or techniques employed to solve the problem(s)

c. Date problem(s) solved

d. Notebook reference

e. Person(s) who solved the problem(s)

f. Person(s) who witnessed and understood the work being done

3B. The fermentation broth contained (check no. 1 or no. 2):

[ ] **1.** One desired compound

[ ] **2.** More than one desired and/or undesired compound

If no. 2 is checked, give

a. Date papergrams show separate components

b. Notebook reference

c. Person(s) who did the work

d. Person(s) who witnessed and understood the work being done

Form 4. Fermentation Procedures

4A. Give a detailed example of the best mode of a complete fermentation procedure. This would include the culture.

In giving this example:

1. Show inoculum size
2. Give vessel size and operating volume
3. Indicate composition of trademark material; also give the generic name of the nutrient and the supplier
4. Define all assay procedures employed in a way that a person skilled in the art would understand. In-house jargon should be avoided.

   In addition to the above, the various other details such as temperature, aeration, agitation, etc. used should be set out.

Form 5. Recovery Process

5A. Give a detailed example of the *best mode* of a complete recovery procedure from whole broth to crude concentrate.

In giving this example:

1. Indicate composition of trademark material; also, give the generic name of the material and the supplier
2. Define all assay procedures employed

5B. Provide a tabulated profile sheet for this best mode of recovery process, if available. *Production* data need not be given *unless* precise figures are necessary to prove the patentability of the invention. Generally, ratios of results are sufficient. The patent attorney should be consulted on this if there is any doubt.

Form 6. Alternate Recovery Process

6A. In general terms, give alternate methods of recovery that *may* be employed. (These need not have been done but rather are deemed *possible* because of the nature of the fermentation product[s].) This information would embody use of various absorption materials, solvent extraction procedures, etc.

Form 7. Recovery Process References

7A. Name person who set up the *initial* successful recovery procedures.

1. Date
2. Notebook reference
3. Person(s) who witnessed and understood the work being done

7B. If more than one desired compound was found in the recovery process:

1. Indicate who first successfully set up a procedure to recover separate compounds
2. Give date
3. Give notebook reference
4. Name person(s) who witnessed and understood the work being done

Form 8. Purification Procedures

8A. Give a detailed example of the best mode of a complete purification procedure from crude concentrate to the final preparation of the desired product.
In giving this example:

1. Indicate composition of trademark material; also, give the generic name of the material and the supplier.
2. Define all assay procedures employed in a way that a person skilled in the art would understand. In-house jargon should be avoided.

Form 9. Alternate Purification Procedures

9A. In general terms, give alternate methods of purification that *may* be employed. (These need not have been done but rather are deemed *possible* because of the nature of the fermentation product[s]).

Form 10. Purification Procedures References

10A. Name person who set up the *initial* successful purification procedure.

1. Date

2. Notebook reference

3. Person(s) who witnessed and understood the work being done

10B. If more than one desired compound was found in the purification process:

1. Indicate who first successfully set up a procedure to recover separate compounds

2. Give date

3. Give notebook reference

4. Name person(s) who witnessed and understood the work being done

Form 11. Salts

11A. List the salts of the desired compound(s) made and the procedure employed.

11B. Indicate who set up the procedure for making the salts.

11C. Give date.

11D. Give notebook reference.

11E. Name person(s) who witnessed and understood the work being done.

11F. List the salts which *can* be made in view of the known characteristics of the compound. (These need not have been made.)

Form 12. Derivatives

12A. List the derivatives of the desired compound(s) made and the procedure employed.

12B. Indicate who set up the procedure for making the derivatives.

12C. Give date.

12D. Give notebook reference.

12E. Name person(s) who witnessed and understood the work being done.

12F. List the derivatives that *can* be made in view of the known characteristics of the compound. (These need not have been made.)

Form 13. Characterization

13A. The fermentation product is (check pertinent number):

[ ] **1.** Acidic
[ ] **2.** Basic
[ ] **3.** Amphoteric
[ ] **4.** Neutral

13B. Check the characterization data obtained on the compound(s) and identify the preparation by the notebook reference (check pertinent numbers):

[ ] **1.** IR curve and tabulation
Reference

[ ] **2.** UV absorption spectrum
Reference

[ ] **3.** Titration data
Reference

[ ] **4.** Elemental analysis
Reference

[ ] **5.** Molecular weight
Reference

[ ] **6.** Optical rotation
Reference

[ ] **7.** Solubilities
Reference

[ ] **8.** Antibacterial spectrum
Reference

[ ] **9.** Antifungal spectrum
Reference

[ ] **10.** Antiviral activity
Reference

[ ] **11.** Papergrams
Reference

[ ] **12.** In vivo activity
Reference

[ ] **13.** Other characterization data not listed above
Reference

[ ] **14.** Differentiating characteristics over prior art, if available
Reference

Append to this form all characterization data checked above.

Negatives of the IR curve should be retained in a safe place.

Form 14. Utility

14A. List the *non-therapeutic* uses for the compound(s):

14B. List the *therapeutic* uses, including veterinary, for the compound(s):

14C. The derivatives and/or salts of the compound(s) can be used for the following *non-therapeutic* uses:

14D. The derivatives and/or salts of the compound(s) can be used for the following *therapeutic* uses, including veterinary:

14E. Give *details* as to how the invention can be used for the above-stated purposes. References to applicable published procedures can supplement the disclosure.

Form 15. Patent and Literature References

15A. List all *known* patent and literature references that are related to the fermentation product(s) and/or process for preparing the same. Submit copies of these references with this Invention Disclosure to the patent attorney.

Form 16. Prior Disclosure

16A. List *all* previous disclosures of this invention. In making this list, give the audience, date, and means of disclosure.

16B. List *all* previous uses of this invention. In making this list, give the date and details of the use.

2. For a Genetic Engineering Invention

If the biotechnology invention is specifically concerned with a genetic engineering invention, a set of *forms* which can be used for the invention disclosure is as follows:

Form 1. General Description of the Invention

1A. Give a brief description of the invention (no more than one page).

1B. Append a schematic sketch of the *process* used to make the invention.

1C. Identify the person(s) who conceived the invention and the person(s) who did the laboratory procedure.

1D. List the notebook(s) and identify the pages disclosing the subject invention.

Form 2. Details of the Invention

2A. Give a detailed description of the best mode contemplated for practicing the invention. For example, if the invention is the construction of a hybrid plasmid, detail the best procedure known for doing the construction.

2B. List all *critical* features of the *process* to make the invention.

2C. List the *alternative procedures* that can be used in the process to make the invention.

2D. List all trademark materials used in the process to make the invention and identify their nature and source.

2E. When certain *steps* in the *process* are *standard,* i.e., well known in the art at the time the invention was made, identify such steps by stating that they are standard procedures. Also, disclose a publication which contains a description of the *standard step(s).*

Form 3. Microorganisms

3A. List all microorganisms used as DNA sources or host microbes in making the invention. In giving this list, record known depository accession numbers, and/or commercial sources of such microbes.

Form 4. Vectors

4A. When a *known* plasmid or bacteriophage is used in the invention, give a reference to a publication that characterizes the same by a restriction endonuclease cleavage map.

4B. Give culture repository accession numbers for all *known* vectors used in the invention. If the vector(s) are available from a commercial source, identify the source.

4C. When a *novel* plasmid or bacteriophage is used in the invention, append a restriction endonuclease cleavage map of the same to this form.

Form 5. DNA

5A. Identify the *source* of any DNA sequence used in the invention.

5B. Give the nucleotide sequence of any DNA sequence used in the invention.

Form 6. Genes

6A. Identify the natural source of any gene used in the invention.

6B. *Detail the procedure* for isolating and purifying the gene from its natural source.

6C. If the gene is synthesized, give the details of synthesis.

6D. Characterize the gene by the best available art-recognized data. If the complete or partial nucleotide sequence is known, detail this data.

6E. Identify all regulatory and/or control regions surrounding the gene.

Form 7. Utility

7A. Identify the use(s) for the invention.

7B. Detail how the invention can be used for the above-stated purposes. Reference to published procedures in which the invention can be used can be listed.

Form 8. Patent and Literature References

8A. List all known patent and literature references that are related to the genetic engineering product(s) and/or process for preparing the same. Submit copies of these references with this Invention Disclosure.

Form 9. Prior Disclosure

9A. List *all* previous disclosures of this invention. In making this list, give the audience, date, and means of disclosure.

9B. List *all* previous uses of this invention. In making this list, give the date and details of such use(s).

Appendix B

Culture Depositories and Microorganisms Accepted

Culture depositories that have been approved as suitable for deposits under the Budapest Treaty provisions—and the microorganisms accepted by each depository—are listed below in alphabetical order by country.

Bulgaria

National Bank for Industrial Microorganisms and Cell Cultures (NBIMCC)
125, Lenin Blvd.
Block 2
Sofia
Bulgaria

> Bacteria, actinomycetes, microscopic fungi, yeasts, microscopic algae, animal cell lines, animal viruses and plasmid containing microorganisms

France

Collection Nationale de Cultures de Micro-organismes (CNCM)

Institut Pasteur
28 Rue du Dr Roux
75724 Paris Cedex 15
France

Bacteria (including actinomycetes), bacteria containing plasmids; filamentous fungi and yeasts, and viruses, EXCEPT:

- cellular cultures (animal cells, including hybridomas and plant cells)
- microorganisms whose manipulation calls for physical insulation standards of P3 or P4 level, according to the information provided by the National Institutes of Health (NIH) Guidelines for Research Involving Recombinant DNA Molecules and Laboratory Safety Monograph
- microorganisms liable to require viability testing that the CNCM is technically not able to carry out
- mixtures of undefined and/or unidentifiable microorganisms

The CNCM reserves the possibility of refusing any microorganism for security reasons; specific risks to human beings, animals, plants and the environment.

In the eventuality of the deposit of cultures that are not or cannot be lyophilized, the CNCM must be consulted, prior to the transmittal of the microorganism, regarding the possibilities and conditions for acceptance of the samples; however, it is advisable to make this prior consultation in all cases.

Germany

Deutsche Sammlung von Mikroorganismen (DSM)

Gesellschaft für Biotechnologische
Forschung mbH
Grisebachstr. 8
3400 Göttingen
Federal Republic of Germany

Bacteria, including actinomycetes, fungi, including yeasts, bacteriophages, except any kind pathogenic to humans or animals. Phytopathogenic kinds are accepted, EXCEPT:

— *Erwinia amylovora; Coniothyrium fagacearum; Endothia parasitica; Gloeosporium ampelophagum; Septoria musiva; Synchytrium endobioticum.*

Hungary

National Collection of Agricultural and Industrial Microorganisms (NCAIM)

Department of Microbiology
University of Horticulture
Somlói ut 14-16
H-1118 Budapest
Hungary

Bacteria (including Streptomyces) EXCEPT:

— obligate human pathogenic species (e.g., *Corynebacterium diphtheriae, Mycobacterium leprae, Yersinia pestis,* etc.)

Fungi, including yeasts and molds, EXCEPT:

— some pathogens (Blastomyces, Coccidioides, Histoplasma, etc.) as well as certain basidiomycetous and plant pathogenic fungi which cannot be preserved reliably.

Apart from the above-mentioned, the following may not, at present, be accepted for deposit:

— viruses, phages, rickettsiae

— algae, protozoa

— cell lines, hybridomas

Japan

Fermentation Research Institute (FRI)

1-3, Higashi 1-chome
Yatabe-machi
Tsukuba-gun, Ibaraki-ken 305
Japan

Fungi, yeast, bacteria and actinomycetes, EXCEPT:

— microorganisms having properties which are or may be dangerous to health or the environment

— microorganisms which need the physical containment level P2, P3 or P4 required for experiments, as described in the 1979 Prime Minister's Guideline for Research Involving Recombinant DNA Molecules.

Netherlands

Centraalbureau voor Schimmelcultures (CBS)
Oosterstraat 1
Postbus 273
NL-3740 AG Baarn
Netherlands

Fungi, including yeasts; actinomycetes, bacteria other than actinomycetes.

United Kingdom

Culture Collection of Algae and Protozoa (CCAP)
Freshwater Biological Association
Windermere Laboratory
The Ferry House
Far Sawrey
Ambleside, Cumbria LA22 0LP
United Kingdom

Freshwater and terrestrial algae and free-living protozoa.

Culture Collection of Algae and Protozoa (CCAP)
Scottish Marine Biological Association
Dunstaffnage Marine Research Laboratory
P.O. Box 3
Oban, Argyll PA34 4AD
United Kingdom

Marine algae, other than large seaweeds.

Culture Collection of the Commonwealth Mycological Institute (CMI CC)
Ferry Lane
Kew, Surrey TW9 3AF
United Kingdom

Fungal isolates, other than known human and animal pathogens and yeasts, that can be preserved without significant change to their properties by the methods of preservation in use.

European Collection of Animal Cell Cultures (ECACC)

Vaccine Research and Production Laboratory
Public Health Laboratory Service
Centre for Applied Microbiology and Research
Porton Down
Salisbury, Wiltshire SP4 0JG
United Kingdom

Cell lines that can be preserved without significant change to or loss of their properties by freezing and long term storage; viruses capable of assay in tissue culture. A statement on their possible pathogenicity to man and/or animals is required at the time of deposit. Up to and including ACDP Category 3 (Advisory Committee on Dangerous Pathogens: Categorisation of Pathogens according to Hazard and Categories of Containment ISBN 0/11/883761/3 HMSO London) can be accepted for deposit.

National Collection of Industrial Bacteria (NCIB)

c/o The National Collection of Industrial and Marine Bacteria Ltd.
Torry Research Station
P.O. Box 31
135 Abbey Road
Aberdeen AB9 8DG
United Kingdom

(a) Bacteria, including actinomycetes, that can be preserved without significant change to their properties by liquid nitrogen freezing or by freeze-drying (lyophilization), and which are allocated to a hazard group no higher than Group 2 as defined by the UK Advisory Committee on Dangerous Pathogens (ACDP).
(b) Plasmids, including recombinants, either (i) cloned into a bacterial or actinomycete host, or (ii) as naked DNA preparations. As regards (i) above, the hazard category of the host with or without its plasmid must be no higher than ACDP Group 2.
As regards (ii) above, the phenotypic markers of the plasmid must be capable of expression in a bacterial or actinomycete host and must be

readily detectable. In all cases, the physical containment requirements must not be higher than level II as defined by the UK Genetic Manipulation Advisory Group (GMAG) and the properties of the deposited material must not be changed significantly by liquid nitrogen freezing or freeze-drying.

(c) Bacteriophages that have a hazard rating and containment requirement no greater than those cited in (a) or (b), above, and which can be preserved without significant change to their properties by liquid nitrogen freezing or by freeze drying.

Notwithstanding the foregoing, the NCIB reserves the right to refuse to accept any material for deposit which in the opinion of the Curator presents an unacceptable hazard or is technically too difficult to handle.

In exceptional circumstances the NCIB may accept deposits which can only be maintained in active culture, but acceptance of such deposits, and relevant fees, must be decided on an individual basis by prior negotiation with the prospective depositor.

National Collection of Type Cultures (NCTC)

Central Public Health Laboratory
61 Colindale Avenue
London NW9 5HT
United Kingdom

Bacteria that can be preserved without significant change to their properties by freeze-drying and which are pathogenic to man and/or animals.

National Collection of Yeast Cultures (NCYC)

Food Research Institute
Colney Lane
Norwich, Norfolk NR4 7UA
United Kingdom

Yeasts other than known pathogens that can be preserved without significant change to their properties by freeze-drying or, exceptionally, in active culture.

United States

American Type Culture Collection (ATCC)

12301 Parklawn Drive
Rockville, Maryland 20852 USA

Following is a list of types of material accepted for deposit by the ATCC:
Algae, animal and plant viruses, bacteria, bacteriophages, fungi, plant tissue cultures, plasmids, protozoa, seeds (25 per sample) and yeasts (except Preceptrol cultures).

The ATCC must be informed of the physical containment level required for experiments using the host vector system, as described in the 1980 National Institutes of Health Guidelines for Research Involving Recombinant DNA Molecules (i.e., P1, P2, P3 or P4 facility). The ATCC, for the time being, will accept only those hosts containing plasmids which can be used in a P1 or P2 facility.

Certain animal viruses may require viability testing in an animal host, which the ATCC may be unable to provide. In such case, the deposit cannot be accepted. Plant viruses which cannot be mechanically inoculated also cannot be accepted.

In Vitro International, Inc. (IVI)

611 (P) Hammonds Ferry Road
Linthicum, Maryland 21090
United States of America

Algae, bacteria with plasmids, bacteriophages, cell cultures, fungi, protozoa and animal and plant viruses. Recombinant strains of microorganisms will also be accepted, but IVI must be notified in advance of accepting the deposit of the physical containment level required for the host vector system, as prescribed by the National Institutes of Health Guidelines. At present, IVI will accept only hosts containing recombinant plasmids that can be used in a P1 or P2 facility.

ARS Patent Collection (NRRL)

Culture Collection Research-Fermentation Laboratory
Northern Regional Research Center
U.S. Department of Agriculture
1815 North University Street
Peoria, Illinois 61604 USA

The types of microorganisms that may be deposited in NRRL are as follows:

Progeny of strains of agriculturally and industrially important bacteria, yeast, molds, and Actinomycetales.

However, the list of microorganisms which may *not* be deposited is much longer and includes the following ten groups:

- *Actinobacillus* (all species); *Actinomyces* (anaerobic/microaerophilic—all species); *Arizona* (all species); *Bacillus anthracis*; *Bartonella* (all species); *Bordetella* (all species); *Borellia* (all species); *Brucella* (all species); *Clostridium botulinum; Clostridium chauvoei; Clostridium haemolyticum; Clostridium histolyticum; Clostridium novyi; Clostridium septicum; Clostridium tetani; Corynebacterium diphtheriae; Corynebacterium equi; Corynebacterium haemolyticum; Corynebacterium pseudotuberculosis; Corynebacterium pyogenes; Corynebacterium renale; Diplococcus* (all species); *Erysipelothrix* (all species); *Escherichia coli* (all enteropathogenic types); *Francisella* (all species); *Haemophilus* (all species); *Herellea* (all species); *Klebsiella* (all species); *Leptospira* (all species); *Listeria* (all species); *Mima* (all species); *Moraxella* (all species); *Mycobacterium avium; Mycobacterium bovis; Mycobacterium tuberculosis; Mycoplasma* (all species); *Neisseria* (all species); *Pasteurella* (all species); *Pseudomonas pseudomallei; Salmonella* (all species); *Shigella* (all species); *Sphaerophorus* (all species); *Staphylococcus aureus; Streptobacillus* (all species); *Streptococcus* (all pathogenic species); *Treponema* (all species); *Vibrio* (all species); *Yersinia* (all species).
- *Blastomyces* (all species); *Coccidioides* (all species); *Cryptococcus* (all species); *Histoplasma* (all species); *Paracoccidioides* (all species).
- *Basidiomycetes* or other molds that cannot successfully be preserved by lyophilization (freeze-drying).
- All viral, rickettsial, and chlamydial agents.
- Agents which may introduce or disseminate any contagious or infectious disease of animals, humans, or poultry and which would require a permit for entry and/or distribution within the United States of America.
- Agents which are classified as Plant Pests and which would require a permit for entry and/or distribution within the United States of America.
- Mixtures of microorganisms.
- Fastidious microorganisms which would require (in the view of the Curator) more than reasonable attention in handling and preparation of lyophilized material.
- Phages of any kind.
- Plasmids and like materials.

USSR

USSR Research Institute for Genetics and Industrial Microorganism Breeding

Dorozhnaya Street, No. 8
113545, Moscow
USSR

Bacteria (including actinomycetes) and microscopic fungi (including yeasts) for essentially industrial and non-medical purposes are accepted for deposit, to the exclusion of microorganisms that cause disease in man and animals and microorganisms that have a toxicogenic effect on plants or require them to be quarantined.

Institute of Microorganism Biochemistry and Physiology of the USSR Academy of Science

Pushchino-na-Oke
142292, Moscow Region
USSR

Bacteria (including actinomycetes) and microscopic fungi (including yeasts), also if they are carriers of recombinant DNA, are accepted for deposit, to the exclusion of microorganisms that cause diseases in man and animals, and microorganisms that have a toxicogenic effect on plants or require them to be quarantined.

USSR Research Institute for Antibiotics

Nagatinskaya Street 3-a
113105, Moscow
USSR

Bacteria (including actinomycetes) and microscopic fungi (including yeasts) for essentially medical purposes are accepted for deposit, to the exclusion of microorganisms that cause disease in man and animals and microorganisms that have a toxicogenic effect on plants or require them to be quarantined.

Appendix C

Subjects of Some Recent Patents

U.S. Patent No.	Title
3,814,796	Antibiotics Zorbonomycin and Zorbonomycin B and Process for the Preparation Thereof
4,211,841	Process for Microbial Transformation of Steroids
4,237,224	Process for Producing Biologically Functional Chimeras
4,246,348	Microbiologic Conversion of L-Galactonate into 2-Keto L-Galactonate
4,262,090	Interferon Production
4,263,404	Rifamycin R and Microorganisms Useful in the Production of Rifamycins
4,264,587	Vaccine for Preventing Persistent Feline Leukemia Viremia in Cats
4,264,588	Vaccine for *Clostridium perfringens* Type E Enterotoxemia of Rabbits
4,264,731	DNA Joining Method
4,264,737	Method of Making Genetically Stable Mutants of *Vibrio cholerae*
4,266,035	Process for Decolorizing Pulp and Paper Mill Wastewater and Microorganisms Capable of Same

U.S. Patent No.	Title
4,267,274	*Streptomyces mediterranei* Mutant Capable of Production of Rifamycin B
4,273,872	Method for the Microbiologic Conversion of Steroids
4,275,155	Manufacture of Aminoglycoside Antibiotics
4,275,157	Method for the Production of L-Lysine
4,275,158	Manufacture of Fatty Acids Having Straight and Long Carbon Chains Using a Microorganism
4,275,163	Cellulase-Producing Microorganism
4,277,466	Complexes of DNA and Esters Derived from Daunorubicine, Their Preparation and Use
4,278,766	Mutant Microorganisms Useful for the Production of Single Cell Protein and Amino Acids
4,282,327	Biologically Pure Culture of *Paecilomyces abruptus*
4,283,489	Purification of Nucleotide Sequences Suitable for Expression in Bacteria
4,283,494	Microbial Lipase, Process for Its Preparation and Microbiologically Pure Culture Therefor
4,293,652	Method for Synthesizing DNA Sequentially
4,294,846	Hypocholesteremic Fermentation Products and Products of Preparation
4,443,539	Process for Preparing Bovine Growth Hormone
4,546,096	Anticholinergic Glucuronides and Antiperspirant Use Thereof
4,574,083	Isolates of Pythium Species Which are Antagonistic to *Pythium ultimum*
4,588,686	Enzymatic Process for Preparing Glucuronides
4,629,792	Chromatographic Separation of Tropicamide O-β-D-Glucuronide Diastereomers
4,643,756	Bioherbicide for Florida Beggarweed
4,687,741	Novel Enzymes Which Catalyze the Degradation and Modification of Lignin
4,691,009	Hybrid Proteins Produced by an Ultrahigh Prokaryotic Expression System
4,695,462	Cellular Encapsulation of Biological Pesticides
4,717,564	High Titer Varicella-Zoster Immune Globulin for Intravenous Administration
4,717,653	Method for Identifying and Characterizing Organisms
4,717,660	Detection of Bacteria by Fluorescent Staining in an Expanded Buffy Coat
4,717,670	Porcine Beta FSH

U.S. Patent No.	Title
4,717,717	Stabilized Compositions Containing Epidermal Growth Factor
4,719,179	Six Base Oligonucleotide Linkers and Methods for Their Use
4,720,386	Vaccine and Method for Immunotherapy of Neoplastic Disease
4,720,454	Genetic Probe Used in the Detection of Adrenal Hyperplasia
4,720,462	Culture System for the Culture of Solid Tissue Masses and Method of Using the Same
4,720,482	Pharmaceutical Compositions Containing One or More Lymphocyte Growth Factors
4,720,483	Oligopeptides and Intermediates and Processes for Their Manufacture
4,720,489	Hair Growth Modification with Ornithine Decarboxylase Inhibitors
4,720,579	Separation of Citric Acid from Fermentation Broth with a Neutral Polymeric Adsorbent
4,721,585	Process for the Biological Treatment of Sewage
4,721,617	Vaccine Against Lyme Disease
4,721,668	Pre-Electrophoretic Nucleic Acid Hybridization
4,721,671	Efficient Prokaryotic Expression System Using Portions of the *E. coli* β-Glucuronidase Gene
4,721,672	cDNA and Gene for Human Angiogenin (Angiogenesis Factor) and Methods of Expression
4,721,673	Recovery and Activation Process for Microbially Produced Calf Prochymosin
4,721,675	Production of Hepatitis A Virus in Vitro Utilizing a Persistently Virus Infected Cell Culture System
4,721,676	Novel Thermophile Isolate Having Thermostable Hydrolytic Activity
4,721,677	Implantable Gas-Containing Biosensor and Method for Measuring an Analyte Such as Glucose
4,721,679	Tissue Typing Tray
4,721,681	Immunoassay in Centrifugal Field with Complementary Particles of Differing Specific Gravities
4,721,730	Adsorbent for β_2-Microglobulin and Immunoglobulin L-Chain
4,721,775	Effective Peptides Related to the Luteinizing Hormone Releasing Hormone from L-Amino Acids

Appendix D

Selected Statutory Patent Law Provisions

The statutory provisions upon which American patent law practice is based are found in Title 35 of the United States Code. All of these provisions have equal force as law, and even relatively obscure provisions may have tremendous significance in any given case. Therefore, a patent attorney must be familiar with all of the statutory provisions and must pass an examination on these and other aspects of patent law before being licensed to practice before the patent office.

Although all of the provisions of Title 35 are important, several provisions play particularly critical roles in patent jurisprudence. Four of the most important provisions are reproduced below; they are numbered as follows: 35 USC 101, 35 USC 102, 35 USC 103, and 35 USC 112. Much of the material that has been discussed in this book pertains directly to these sections and to their legal interpretations.

§101 Inventions patentable

Whoever invents or discovers any new and useful process, machine, manufacture, or composition of matter, or any new and useful improvement thereof, may obtain a patent therefor, subject to the conditions and requirements of this title.

§102 Conditions for patentability; novelty and loss of right to patent

A person shall be entitled to a patent unless:

a. the invention was known or used by others in this country, or patented or described in a printed publication in this or a foreign country, before the invention thereof by the applicant for patent, or
b. the invention was patented or described in a printed publication in this or a foreign country or in public use or on sale in this country, more than one year prior to the date of the application for patent in the United States, or
c. he has abandoned the invention, or
d. the invention was first patented or caused to be patented, or was the subject of an inventor's certificate, by the applicant or his legal representatives or assigns in a foreign country prior to the date of the application for patent in this country on an application for patent or inventor's certificate filed more than twelve months before the filing of the application in the United States, or

e. the invention was described in a patent granted on an application for patent by another filed in the United States before the invention thereof by the applicant for patent, or on an international application by another who has fulfilled the requirements of paragraphs (1), (2), and (4) of section 371(C) of this title before the invention thereof by the applicant for patent, or
f. he did not himself invent the subject matter sought to be patented, or
g. before the applicant's invention thereof the invention was made in this country by another who had not abandoned, suppressed, or concealed it. In determining priority of invention there shall be considered not only the respective dates of conception and reduction to practice of the invention, but also the reasonable diligence of one who was first to conceive and last to reduce to practice, from a time prior to conception by the other. (Amended July 28, 1972, Public Law 92–358, sec. 2, 86 Stat. 501; November 14, 1975, Public Law 94–131, sec. 5, 89 Stat. 691.)

§103 Conditions for patentability; non-obvious subject matter

A patent may not be obtained though the invention is not identically disclosed or described as set forth in section 102 of this title, if the differences between the subject matter sought to be patented and the prior art are such that the subject matter as a whole would have been obvious at the time the invention was made to a person having ordinary skill in the art to which said subject matter pertains. Patentability shall not be negatived by the manner in which the invention was made.

Subject matter developed by another person, which qualifies as prior art only under subsection (f) or (g) of section 102 of this title, shall not preclude patentability under this section where the subject matter and the claimed invention were, at the time the invention was made, owned by the same person or subject to an obligation of assignment to the same person. (Added November 8, 1984, Public Law 98–622, sec. 103, 98 Stat. 3384.)

§112 Specification

The specification shall contain a written description of the invention, and of the manner and process of making and using it, in such full, clear, concise, and exact terms as to enable any person skilled in the art to which it pertains, or with which it is most nearly connected, to make and use the same, and shall set forth the best mode contemplated by the inventor of carrying out his invention.

The specification shall conclude with one or more claims particularly pointing out and distinctly claiming the subject matter which the applicant regards as his invention.

A claim may be written in independent or, if the nature of the case admits, in dependent or multiple dependent form.

Subject to the following paragraph, a claim in dependent form shall contain a reference to a claim previously set forth and then specify a further limitation of the subject matter claimed. A claim in dependent form shall be construed to incorporate by reference all the limitations of the claim to which it refers.

A claim in multiple dependent form shall contain a reference, in the alternative only, to more than one claim previously set forth and then specify a further limitation of the subject matter claimed. A multiple dependent claim shall not serve as a basis for any other multiple dependent claim. A multiple dependent claim shall be construed to incorporate by reference all the limitations of the particular claim in relation to which it is being considered.

An element in a claim for a combination may be expressed as a means or step for performing a specified function without the recital of structure, material, or acts in support thereof, and such claim shall be construed to cover the corresponding structure, material, or acts described in the specification and equivalents thereof. (Amended July 24, 1965, Public Law 89–83, sec. 9, 79 Stat. 261; November 14, 1975, Public Law 94–131, sec. 7, 89 Stat. 691.)

Appendix E

Three Biotechnology Patents

This appendix contains the complete text of three recent U.S. patents in biotechnology. The first patent, which is for a microbiological invention, begins on the next page. The second one is for a bioherbicide and begins on page 155. The third patent, for a genetic engineering invention, begins on page 159.

Copies of all U.S. patents can be obtained from the Commissioner of Patents and Trademarks, Washington, D.C. 20231, for a charge of $1.50 per copy. Appendix C lists the numbers and titles of many recent patents for biotechnology.

United States Patent [19]

Barnes et al.

[11] **Patent Number: 4,695,455**

[45] **Date of Patent: * Sep. 22, 1987**

[54] **CELLULAR ENCAPSULATION OF PESTICIDES PRODUCED BY EXPRESSION OF HETEROLOGOUS GENES**

[75] Inventors: **Andrew C. Barnes,** San Diego; **Susan G. Cummings,** Chula Vista, both of Calif.

[73] Assignee: **Mycogen Corporation,** San Diego, Calif.

[*] Notice: The portion of the term of this patent subsequent to Sep. 22, 2004 has been disclaimed.

[21] Appl. No.: **771,313**

[22] Filed: **Aug. 30, 1985**

Related U.S. Application Data

[63] Continuation-in-part of Ser. No. 693,080, Jan. 22, 1985, abandoned.

[51] **Int. Cl.**[4] **A01N 63/00;** C12P 21/00; C12N 15/00; C12N 1/20

[52] **U.S. Cl.** **424/93;** 435/68; 435/172.3; 435/253; 435/254; 435/317; 435/255; 435/260; 47/58

[58] **Field of Search** 435/68, 172.3, 31, 32, 435/317, 253, 243, 257, 260, 261; 47/58; 424/93, 88, 92

[56] **References Cited**

U.S. PATENT DOCUMENTS

3,779,869	12/1973	Zienty	195/68
4,000,258	12/1976	Shieh et al.	424/93
4,265,880	5/1981	Spence et al.	424/93
4,328,203	5/1982	Spence et al.	424/16
4,337,313	6/1982	Hershberger et al.	435/177
4,374,200	2/1983	Olsen	435/172
4,609,550	9/1986	Fitz-James	424/93

FOREIGN PATENT DOCUMENTS

51-5047	7/1981	Japan	424/93

OTHER PUBLICATIONS

Schnepf et al., *Proc. Natl. Acad. Sci.,* vol. 78 (5) May 1981, "Cloning and Expression of the *Bacillus thuringiensis* Crystal Protein in *Eschenchia coli*", pp. 2893–2897.

Ohgaki et al, *Chem Abst,* vol. 83, 1975, No. 2287m, "Synergistic Antimicrobial Action of Formalin".

Gotterdi et al, *Chem Abst,* vol. 193, No. 80030t, 1980, "Redox Potential and Germicidal Action of Aqueous Halogen Solutions".

Ward et al, *FEBS,* Oct. 1984, vol. 175(2), pp. 377–382, "Cloning and Expression in *Eschenchia coli* of the Insecticidal γ-Endotoxin Gene of *Bacillus thuringiensis,* var. Israelensis".

Hawkes *Electron Optics and Electron Microscopy,* 1972, Taylor & Francis Ltd, pp. 214–215.

Wischitzer, *Introduction to Electron Microscopy,* pp. 125–135, Prergamon Press 1982.

Glauert, Audrey from *Techniques for Electron Microscopy,* 1961.

Fawcett, Don W., from *Modern Developments in Electron Microscopy,* 1964.

Primary Examiner—Thomas G. Wiseman
Assistant Examiner—Robin Lyn Teskin
Attorney, Agent, or Firm—Roman Saliwanchik; David R. Saliwanchik

[57] **ABSTRACT**

Methods and compositions are provided for preparing and using pesticides, where the pesticides are encapsulated in non-proliferating cells. The methods involve introducing a heterologous gene into a cellular host, where expression of the heterologous gene results, directly or indirectly, in production of the pesticide. These cells are then killed under conditions which prolong the pesticidal activity when said cells are applied to the environment of a target pest. The killed cells can be used directly or after formulation for treatment of an agricultural host or environment of the host with the pesticide.

19 Claims, No Drawings

(Continued on next page)

1

CELLULAR ENCAPSULATION OF PESTICIDES PRODUCED BY EXPRESSION OF HETEROLOGOUS GENES

CROSS REFERENCE TO A RELATED APPLICATION

This is a continuation-in-part of our copending application Ser. No. 693,080, now abandoned filed on Jan. 22, 1985.

BACKGROUND OF THE INVENTION

The extraordinary increase in agricultural productivity has been a result of many factors, including significantly better understanding of the methods involved with agriculture, improved equipment, availability of fertilizers, and improved pesticides. The latter factor has not been without detrimental aspects, however, due to the negative effect on the environment. There is, therefore, a substantial interest in developing effective and environmentally acceptable pesticides.

Among ecologically acceptable pesticides are the protein toxins produced by various microorganisms, such as *Bacillus thuringiensis.* However, the use of *B. thuringiensis* lysate or spores as a pesticide has significant drawbacks. The lifetime of the pesticide is relatively short in the environment, requiring multiple applications to give adequate protection. Consequently, these pesticides are not economical in comparison to more traditional chemical products having long residual activities. Improvements in field longevity would greatly aid in expanding the application of biological, or protein toxin-based pesticides.

As indicated above, there are many requirements for pesticides associated with their particular application. For example, in many cases it is desirable to have pesticides which have long residual activity in the field while not accumulating in the environment. In addition, because of economic considerations, it is preferable to have pesticides which have a reasonably broad spectrum of biocidal activity. Also, the pesticide should degrade to degradation products which are environmentally acceptable. Other considerations include ease of formulation, pesticidal activity, stability to environmental effects, such as light, water, organisms, and the like, and effect on beneficial or innocuous organisms in the environment.

U.S. Pat. No. 4,265,880 describes embedding live insecticidal pathogens in a coacervate microbead. Japanese Pat. No. 51-5047 describes physical-chemical methods for killing B.t. spores, while retaining toxicity.

BRIEF SUMMARY OF THE INVENTION

Methods and compositions are provided for protecting agricultural crops and products from pests. The pesticides are produced by introducing a heterologous gene into a cellular host. Expression of the heterologous gene results, directly or indirectly, in the intracellular production and maintenance of the pesticide. These cells are then treated under conditions that prolong the activity of the toxin produced in the cell when the cell is applied to the environment of target pest(s). The resulting product retains the toxicity of the toxin. This naturally encapsulated pesticide may then be formulated in accordance with conventional techniques for application to the environment hosting a target pest, e.g., soil, water, and foliage of plants.

2

DETAILED DISCLOSURE OF THE INVENTION

In accordance with the subject invention, improved pesticides are provided, having among their other advantages an extended residual life, by modifying pesticide producing microorganisms hosting a heterologous gene capable of expression in the host, where expression of the gene results, directly or indirectly, in the production of a pesticide. The subject method involves treating the organisms with reagents that prolong the activity of the toxin produced in the cell when the cell is applied to the environment of target pest(s). The resulting product retains the toxicity of the toxin.

A wide variety of pesticides can be produced which will be characterized by being capable of being produced intracellularly, particularly in a unicellular microorganism host, such as prokaryotes, e.g., bacteria: or eukaryotes, e.g., fungi, exemplified by yeast and filamentous fungi, such as Neurospora and Aspergillus: or protists, such as amoebas, protozoa, algae, and the like.

The pesticide can be any toxin produced by a microbe. For example, it can be a polypeptide which has toxic activity toward a eukaryotic multicellular pest, such as insects, e.g., coleoptera, lepidoptera, diptera, hemiptera, dermaptera, and orthoptera: or arachnids: gastropods: or worms, such as nematodes and platyhelminths. Various susceptible insects include beetles, moths, flies, grasshoppers, lice, and earwigs.

The pesticide which is produced in the host cell may be a polypeptide produced in active form or a precursor or proform which requires further processing for toxin activity, e.g., by the pest, as with the crystal toxin of *B. thuringiensis* var. *kurstaki.* Thus, the gene may encode an enzyme which modifies a metabolite to produce a pesticidal composition.

Among naturally-occurring toxins are the polypeptide crystal toxins of *B. thuringiensis* var. *kurstaki,* active against lepidoptera; *B.t.* var. *israelensis,* active against mosquitoes: *B.t.* M-7, active against coleoptera: *B. thuringiensis* var. *aizawai,* active against spodoptera: and *B. sphaericus,* active against mosquito larvae. Other toxins include those of entomopathogenic fungi, such as beauverin of *Beauveria bassiana* and destruxins of *Metarhizium* spp.; or the broad spectrum insecticidal compounds, such as the avermectins of *Streptomyces avermitilus.* Cultures exemplifying the above are as follows:

Baccillus thuringiensis var. *kurstaki* HD—1—NRRL B-3792 disclosed in U.S. Pat. 4,448,885
Bacillus thuringiensis var. *israelensis*—A.'CC 35646
Bacillus thuringiensis M—7—NRRL B-15939

The following *B. thuringiensis* cultures are available from the U.S. Department of Agriculture (USDA) at Brownsville, Tex. Requests should be made to Joe Garcia, USDA, ARS, Cotton Insects Research Unit, P.O. Box 1033, Brownsville, Tex. 78520 USA.

B. thuringiensis HD2
B. thuringiensis var. *finitimus* HD3
B. thuringiensis var. *alesti* HD4
B. thuringiensis var. *kurstaki* HD73
B. thuringiensis var. *sotto* HD770
B. thuringiensis var. *dendrolimus* HD7
B. thuringiensis var. *kenyae* HD5
B. thuringiensis var. *galleriae* HD29
B. thuringiensis var. *canadensis* HD224
B. thuringiensis var. *entomocidus* HD9
B. thuringiensis var. *subtoxicus* HD109
B. thuringiensis var. *aizawai* HD11
B. thuringiensis var. *morrisoni* HD12

B. thuringiensis var. *ostriniae* HD501
B. thuringiensis var. *tolworthi* HD537
B. thuringiensis var. *darmstadiensis* HD146
B. thuringiensis var. *toumanoffi* HD201
B. thuringiensis var. *kyushuensis* HD541
B. thuringiensis var. *thompsoni* HD542
B. thuringiensis var. *pakistani* HD395
B. thuringiensis var. *israelensis* HD567
B. thuringiensis var. *indiana* HD521
B. thuringiensis var. *dakota*
B. thuringiensis var. *tohokuensis* HD866
B. thuringiensis var. *kumanotoensis* HD867
B. thuringiensis var. *tochigiensis* HD868
B. thuringiensis var. *colmeri* HD847
B. thuringiensis var. *wuhanensis* HD525
Bacillus cereus—ATCC 21281
Bacillus moritai—ATCC 21282
Bacillus popilliae—ATCC 14706
Bacillus lentimorbus—ATCC 14707
Bacillus sphaericus—ATCC 33203
Beauveria bassiana—ATCC 9835
Metarhizium anisopliae—ATCC 24398
Metarhizium flavoviride—ATCC 32969
Streptomyces avermitilus—ATCC 31267

The toxin need not be the same as a naturally occurring toxin. Polypeptide toxins may be fragments of a naturally-occurring toxin; expression products of deletion, transversion or transition mutations, where two or fewer number percent of the amino acids may be changed; or a repetitive sequence capable of processing by the intended pest host. In addition, fusion products may be prepared where one, five or more amino acids are provided at the N-terminus to provide, for example, reduced proteolytic degradation of the toxin(s). In some instances, a plurality of the same or different toxins may be encoded and expressed, where processing sites may be introduced between each toxin moiety in the polytoxin.

Illustrative host cells may include either prokaryotes or eukaryotes, normally being limited to those cells which do not produce substances toxic to higher organisms, such as mammals. However, organisms which produce substances toxic to higher organisms could be used, where the toxin is unstable or the level of application sufficiently low as to avoid any possibility of toxicity to a mammalian host. As hosts, of particular interest will be the prokaryotes and lower eukaryotes, such as fungi. Illustrative prokaryotes, both Gram-negative and -positive, include Enterobacteriaceae, such as Escherichia, Erwinia, Shigella, Salmonella, and Proteus; Bacillaceae; Rhizobiaceae, such as Rhizobium; Spirillaceae, such as photobacterium, Zymomonas, Serratia, Aeromonas, Vibrio, Desulfovibrio, Spirillum; Lactobacillaceae; Pseudomonadaceae, such as Pseudomonas and Acetobacter; Azotobacteraceae and Nitrobacteraceae. Among eukaryotes are fungi, such as Phycomycetes and Ascomycetes, which includes yeast, such as Saccharomyces and Schizosaccharomyces; and Basidiomycetes yeast, such as Rhodotorula, Aureobasidium, Sporobolomyces, and the like.

Characteristics of particular interest in selecting a host cell for purposes of production include ease of introducing the heterologous gene into the host, availability of expression systems, efficiency of expression, stability of the pesticide in the host, and the presence of auxiliary genetic capabilities. Characteristics of interest for use as a pesticide microcapsule include protective qualities for the pesticide, such as thick cell walls, pig-

mentation, and intracellular packaging or formation of inclusion bodies; leaf affinity: lack of mammalian toxicity attractiveness to pests for ingestion; ease of killing and fixing without damage to the toxin and the like. Other considerations include ease of formulation and handling, economics, storage stability, and the like.

Host organisms of particular interest include yeast, such as Rhodotorula sp., Aureobasidium sp., Saccharomyces sp., and Sporobolomyces sp.; phylloplane organisms such Pseudomonas sp., Erwinia sp. and Flavobacterium sp., or such other organisms as Escherichia, *Lactobacillus* sp., *Bacillus* sp., and the like. Specific organisms include *Pseudomonas aeruginosa, Pseudomonas fluorescens, Saccharomyces cerevisiae, Bacillus thuringiensis*, Escherichia coli, Bacillus subtilis, and the like.

The cell will usually be intact and be substantially in the proliferative form when killed, rather than in a spore form, although in some instances spores may be employed.

The cells may be inhibited from proliferation in a variety of ways, so long as the technique does not deleteriously affect the properties of the pesticide, nor diminish the cellular capability in protecting the pesticide. The techniques may involve physical treatment, chemical treatment, changing the physical character of the cell or leaving the physical character of the cell substantially intact, or the like.

Various techniques for inactivating the host cells include heat, usually 50° C. to 70° C.; freezing; UV irradiation; lyophilization; toxins, e.g., antibiotics; phenols; anilides, e.g., carbanilide and salicylanilide; hydroxyurea; quaternaries alcohols; antibacterial dyes; EDTA and amidines non-specific organic and inorganic chemicals, such as halogenating agents, e.g., chlorinating, brominating or iodinating agents; aldehydes, e.g., glutaraldehyde or formaldehyde; toxic gases, such as ozone and ethylene oxide peroxide; psoralens; desiccating agents; or the like, which may be used individually or in combination. The choice of agent will depend upon the particular pesticide, the nature of the host cell, the nature of the modification of the cellular structure, such as fixing and preserving the cell wall with crosslinking agents, or the like.

The cells generally will have enhanced structural stability which will enhance resistance to environmental degradation in the field. Where the pesticide is in a proform, the method of inactivation should be selected so as not to inhibit processing of the proform to the mature form of the pesticide by the target pest pathogen. For example, formaldehyde will crosslink proteins and could inhibit processing of the proform of a polypeptide pesticide. The method of inactivation or killing retains at least a substantial portion of the bioavailability or bioactivity of the toxin.

The heterologous gene(s) may be introduced into the host in any convenient manner, either providing for extrachromosomal maintenance or integration into the host genome. (By heterologous is intended that the gene is not present in the host into which it is introduced, nor would the gene normally be found in such host. That is, even if the host organism and the source of the heterologous gene exchange information, the heterologous gene would normally not be found in the wild-type host cells in nature. Usually, the term heterologous will involve species of different genera as host and gene source.)

Various constructs may be used, which include replication systems from plasmids, viruses, or centromeres in combination with an autonomous replicating segment

5

(ars) for stable maintenance. Where only integration is desired, constructs can be used which may provide for replication, and are either transposons or have transposon-like insertion activity or provide for homology with the genome of the host. Frequently, DNA sequences are employed having the heterologous gene between sequences which are homologous with sequences in the genome of the host, either chromosomal or plasmid. Desirably, the heterologous gene(s) will be present in multiple copies. See for example, U.S. Pat. No. 4,399,216. Thus, conjugation, transduction, transfection and transformation may be employed for introduction of the heterologous gene.

A large number of vectors are presently available which depend upon eukaryotic and prokaryotic replication systems, such as ColE1, P-1 incompatibility plasmids, e.g., pRK290, yeast 2m μ plasmid, lambda, and the like.

Where an extrachromosomal element is employed, the DNA construct will desirably include a marker which allows for a selection of those host cells containing the construct. The marker is commonly one which provides for biocide resistance, e.g., antibiotic resistance or heavy metal resistance, complementation providing prototrophy to an auxotrophic host, or the like. The replication systems can provide special properties, such as runaway replication, can involve cos cells, or other special feature.

Where the heterologous gene(s) has transcriptional and translational initiation and termination regulatory signals recognized by the host cell, it will frequently be satisfactory to employ those regulatory features in conjunction with the heterologous gene. However, in those situations where the heterologous gene is modified, as for example, removing a leader sequence or providing a sequence which codes for the mature form of the pesticide, where the entire gene encodes for a precursor, it will frequently be necessary to manipulate the DNA sequence, so that a transcriptional initiation regulatory sequence may be provided which is different from the natural one.

A wide variety of transcriptional initiation sequences exist for a wide variety of hosts. The sequence can provide for constitutive expression of the pesticide or regulated expression, where the regulation may be inducible by a chemical, e.g., a metabolite, by temperature, or by a regulatable repressor. See for example, U.S. Pat. No. 4,374,927. The particular choice of the promoter will depend on a number of factors, the strength of the promoter, the interference of the promoter with the viability of the cells, the effect of regulatory mechanisms endogenous to the cell on the promoter, and the like. A large number of promoters are available from a variety of sources, including commercial sources.

The cellular host containing the heterologous pesticidal gene may be grown in any convenient nutrient medium, where the DNA construct provides a selective advantage, providing for a selective medium so that substantially all or all of the cells retain the heterologous gene. These cells may then be harvested in accordance with conventional ways and modified in the various manners described above. Alternatively, the cells can be fixed prior to harvesting.

The method of treating the host organism containing the toxin can fulfill a number of functions. First, it may enhance structural integrity. Second, it may provide for enhanced proteolytic stability of the toxin, by modifying the toxin so as to reduce its susceptibility to proteolytic degradation and/or by reducing the proteolytic activity of proteases naturally present in the cell. The cells are preferably modified at an intact stage and when there has been a substantial build-up of the toxin protein. These modifications can be achieved in a variety of ways, such as by using chemical reagents having a broad spectrum of chemical reactivity. The intact cells can be combined with a liquid reagent medium containing the chemical reagents, with or without agitation at temperatures in the range of about $-10°$ to $60°$ C. The reaction time may be determined empirically and will vary widely with the reagents and reaction conditions. Cell concentrations will vary from about $10E^2$ to $10E^{10}$ per ml.

6

Of particular interest as chemical reagents are halogenating agents, particularly halogens of atomic no. 17–80. More particularly, iodine can be used under mild conditions and for sufficient time to achieve the desired results. Other suitable techniques include treatment with aldehydes, such as formaldehyde and glutaraldehyde anti-infectives, such as zephiran chloride and cetylpyridinium chloride alcohols, such as isopropyl and ethanol various histologic fixatives, such as Bouin's fixative and Helly's fixative (See: Humason, Gretchen L., Animal Tissue Techniques, W. H. Freeman and Company, 1967); or a combination of physical (heat) and chemical agents that prolong the activity of the toxin produced in the cell when the cell is applied to the environment of the target pest(s).

For halogenation with iodine, temperatures will generally range from about 0° to 50° C., but the reaction can be conveniently carried out at room temperature. Conveniently, the iodination may be performed using triiodide or iodine at 0.5 to 5% in an acidic aqueous medium, particularly an aqueous carboxylic acid solution that may vary from about 0.5–5M. Conveniently, acetic acid may be used, although other carboxylic acids, generally of from about 1 to 4 carbon atoms, may also be employed. The time for the reaction will generally range from less than a minute to about 24 hrs, usually from about 1 to 6 hrs. Any residual iodine may be removed by reaction with a reducing agent, such as dithionite, sodium thiosulfate, or other reducing agent compatible with ultimate usage in the field. In addition, the modified cells may be subjected to further treatment, such as washing to remove all of the reaction medium, isolation in dry form, and formulation with typical stickers, spreaders, and adjuvants generally utilized in agricultural applications, as is well known to those skilled in the art.

Of particular interest are reagents capable of cross-linking the cell wall. A number of reagents are known in the art for this purpose. The treatment should result in enhanced stability of the pesticide. That is, there should be enhanced persistence or residual activity of the pesticide under field conditions. Thus, under conditions where the pesticidal activity of untreated cells diminishes, the activity of treated cells remains for periods of from 1 to 3 times longer.

The cells may be formulated in a variety of ways. They may be employed as wettable powders, granules or dusts, by mixing with various inert materials, such as inorganic minerals (phyllosilicates, carbonates, sulfates, phosphates, and the like) or botanical materials (powdered corncobs, rice hulls, walnut shells, and the like). The formulations may include spreader-sticker adjuvants, stabilizing agents, other pesticidal additives, or surfactants. Liquid formulations may be aqueous-based or non-aqueous and employed as foams, gels, suspensions, emulsifiable concentrates, or the like. The ingredients may include rheological agents, surfactants, emulsifiers, dispersants, or polymers.

The pesticidal concentration will vary widely depending upon the nature of the particular formulation, particularly whether it is a concentrate or to be used directly. The pesticide will be present in at least 1% by weight and may be 100% by weight. The dry formulations will have from about 1–95% by weight of the pesticide while the liquid formulations will generally be from about 1–60% by weight of the solids in the liquid phase. The formulations will generally have from about 1E2 to about 1E4 cells/mg. These formulations will be administered at about 2 oz (liquid or dry) to 2 or more lb/ha.

The formulations can be applied to the environment of the pest(s), e.g., plants, soil or water, by spraying, dusting, sprinkling or the like.

The following examples are offered by way of illustration and not by way of limitation.

EXAMPLE 1 CONSTRUCTION OF A HETEROLOGOUS GENE AND TRANSFORMATION INTO A SUITABLE HOST

A construction began with a clone of *Pseudomonas aeruginosa*, available from Northern Regional Research Laboratories (NRRL B-12127), containing a broad host range shuttle plasmid pRO1614 (J. Bact. [1982]150:60: U.S. Pat. No. 4,374,200). The plasmid has unique HindIII, BamHI, and SalI and PvuII restriction sites, a PstI insertion, which includes the carbenicillin resistance gene and a *P. aeruginosa* replication system, where the HindIII, BamHi and SalI restriction sites are in a tetracycline resistance gene. The remainder of the plasmid is derived from pBR322. A second plasmid, pSM-117, has been deposited as a clone of *E. coli* (NRRL B-15976). This deposit was made with the permanent collection of the Northern Regional Research Laboratory, U.S. Department of Agriculture, Peoria, Ill. 61604, USA. Plasmid pSM-117 confers ampicillin resistance to *E. coli* and contains a 6.8 Kbp HindIII DNA fragment that includes the δ-endotoxin gene from the 50 md plasmid of *Bacillus thuringiensis* HD73. Sufficient toxin is expressed from this gene in *E. coli* to make the intact cells toxic to cabbage loopers. A further modification of the DNA fragment was done to enhance toxin expression and to accomplish expression of toxin in an alternate host, *Pseudomonas fluorescens*. In order to eliminate unwanted DNA from the 6.8 Kbp fragment, pSM-117 was digested with BamHl to open the circular plasmid at a site nearest to the 5′ end of the toxin gene, and was treated with the exonuclease Bal31 to reduce the 6.8 Kbp HindII insert to about 2.9 Kbp. The free ends of the DNA were filled in with Klenow polymerase, BamHl linkers were added, and the linear DNA was ligated to reform the circular plasmid. The resultant vector, pFG270, was digested with XhoI to open the circular plasmid at a site nearest to the 3′ end of the toxin gene. SalI linkers were added, the plasmid was ligated to its circular form, and the resultant vector, pFG10.6 was amplified in *E. coli.* pFG10.6 was digested completely with SalI and BamHI and the resulting 2.1 Kbp fragment containing the toxin gene was purified by gel electrophoresis and ligated into the BamHI and SalI sites of plasmid pRO1614. The resultant plasmid, pCH2.1, was amplified in *E. coli* and used to transform

Pseudomonas fluorescens to carbenicillin resistance and the ability to synthesize δ-endotoxin. *Pseudomonas fluorescens* (pCH2.1) was deposited with NRRL and was given the accession number NRRL B-15977. In the following illustrative experiments *P. fluorescens* (pCH2.1) was used in conjunction with controls of untransformed cells.

The culture deposits *E. coli* (pSM-117)--NRRL B-15976 and *P. fluorescens* (pCH2.1)--NRRL B-15977 were made in the permanent collection of the NRRL repository, to be maintained for at least thirty years. These deposits are available to the public upon the grant of a patent disclosing them. The deposits are also available as required by foreign patent laws in countries wherein counterparts of the subject application, or its progeny, are filed. However, it should be understood that the availability of a deposit does not constitute a license to practice the subject invention in derogation of patent rights granted by governmental action.

EXAMPLE 2 TREATMENT AND TESTING OF MICROBES HOSTING A HETEROLOGOUS GENE

The Pseudomonads were either killed by 1% lugol's iodine (100 g KI, 50 g I_2, 100 ml glacial acetic acid in 1 L sterile distilled water) or by treatment with a 2% formalin solution, each at ambient temperatures.

Cell pellets from two liter broth cultures of *P. fluorescens* with and without *B.t.* toxin are divided in half. Half the cells are treated with 1% lugol's iodine for 4 hr, while the other half are held on ice. Washed lugol's treated cells and untreated cells are repelleted in 10 ml of sterile deionized water. Nine ml of this suspension (10E8-10E12 cells/ml) are sprayed on three young lettuce plants. The three sprayed plants are placed in a single enclosed chamber and 50–75 *Trichoplusia ni* larvae are applied to the plants. Plants are considered protected if there is no visible loss of foliage over the observation period.

TABLE 1

	Plant Assay 1 - *P. fluorescens*			
Microorganism	Treatment	# Viable Cells/ml	Total # Larvae	Protection
P. fluorescens	Live	2.5 × 10E11	75	Not protected
P. fluorescens + Bt Toxin	Live	3.1 × 10E11	75	Protected
P. fluorescens	1% Lugol's Iodine 4 hrs	0	75	Not protected
P. fluorescens + Bt Toxin	1% Lugol's Iodine 4 hrs	0	75	Protected

*Set up day 1, 25 2-day larvae applied to plants, day 3, 50 5-day larvae applied to plants, day 10, observation reported.

In the next study, newly hatched cabbage loopers are placed in a Petri dish containing droplets of the material to be bioassayed. The larvae imbibe the liquid and are then removed to small cups containing larval diet. The larvae are examined after seven days and the total number of animals killed is recorded. The following Table 2 indicates the results.

TABLE 2

Microorganism	Treatment	# Larvae Killed/ Total	% Larvae Killed	Viable Cell Count/ ml
Bioassay 1 - *P. fluorescens*				
P. fluorescens	Live	0/15	0	8 × 10E11
P. fluorescens + Bt Toxin	Live	11/18	62	2.5 × 10E12
P. fluorescens	1% Lugol's 4 hrs	0/15	0	0
P. fluorescens + Bt Toxin	1% Lugol's 4 hrs	8/13	62	0
Bioassay 2 - *P. fluorescens*				
P. fluorescens	Live	0/15	0	3.7 × 10E11
P. fluorescens + Bt Toxin	Live	3/20	15	4.5 × 10E11
P. fluorescens	1% Lugol's 4 hrs	0/15	0	0
P. fluorescens + Bt Toxin	1% Lugol's 4 hrs	8/15	53	0
Bioassay 3 - *P. fluorescens*				
P. fluorescens + Bt Toxin	Live-frozen 14 days	9/10	90	2.5 × 10E12
P. fluorescens	1% Lugol's 4 hrs	1/15	7	0
P. fluorescens + Bt Toxin	1% Lugol's	11/15	73	0
P. fluorescens	2% formalin 4 hrs	0/15	0	0
P. fluorescens + Bt Toxin	2% formalin 4 hrs	10/15	67	0
Bioassay 4 - *P. fluorescens*				
P. fluorescens	Live	6/14	30	3 × 10E11
P. fluorescens + Bt Toxin	Live	20/20	100	3 × 10E10
P. fluorescens	1% Lugol's 4 hrs	0/20	0	0
P. fluorescens + Bt Toxin	1% Lugol's 4 hrs	18/20	90	0

In the next study, different methods of killing the cells were employed to determine the effect on cell stability to sonication. *P. fluorescens* strain 33 is an untransformed cell not containing the Bt toxin, while strain pCH contains the 2.1 kb toxin gene. In the first study, stationary phase cultures of strain 33 and strain pCH were harvested by centrifugation and cell pellets suspended in sterile deionized water at concentrations of 10E10 and 10E9, respectively. Aliquots of cells were exposed to a 70° C. water bath for varying amounts of time and cell viability measured by plate counts on King's B agar. (See King, E. O. et al.(1954) J. Lab. & Clin. Med. 44:304. The medium has the following ingredients: Proteose peptone No. 3, 2.0 percent; Bacto agar, 1.5 percent; glycerol, C.P., 1.0 percent K_2HPO_4 [anhydrous], 0.15 percent: $MgSO_4 \cdot 7H_2O$, 0.15 percent; adjusted to pH 7.2.) Strain 33 was substantially completely killed within 5 min, while strain pCH showed substantially no viable cells within 10 min.

Cells of strain 33 (untreated, heat treated at 70° C., or lugol's treated [1% lugol's iodine, 2 hr, room temperature]) were suspended in 10 ml deionized water to give a cell concentration of approximately 10E9/ml. The cell suspension was then subjected to sonication for 5 min on a Bronson 200 Sonifier (using a microprobe output 10, 50% duty cycle, pulse mode). The optical density (575 nm) of the cell suspension was measured before and after the sonication treatment. A decrease in optical density indicates cell disruption. The following Tables 3 and 4 give the results.

TABLE 3

Sonication - *P. fluorescens* 33		
	Optical Density (575 nm)	
Treatment	Pre-sonication	Post-sonication
Live	0.24	0.05
Lugol's (1%, 2 hr, room temp)	0.20	0.19
Heat 70° C.		
1	0.22	0.06
3	0.21	0.07
5	0.22	0.09
15	0.20	0.10
30	0.21	0.11
60	0.19	0.13
90	0.22	0.13

Cells of strain pCH (5×10E10/ml) were heat treated in a 70° C. water bath and assayed for toxicity against *T. ni* larvae. The bioassay procedure involved newly hatched cabbage loopers which were placed in a petri dish containing droplets of the materials to be bioassayed. The larvae were allowed to imbibe the liquid and were then removed to small cups containing larval diet. The larvae were examined after six days and the total number of animals killed recorded.

TABLE 4

Bioassay Heat Treated *P. fluorescens*			
		# Larvae Killed/Total Larvae	
Treatment (70° C.) Time (min)	# Viable Cells/Total Cells	1×	% Killed
3	2 × 10E5/5.5 × 10E10	8/8	100
5	2 × 10E6/5.5 × 10E10	8/8	100
10	0/5.5 × 10E10	8/8	100

It is evident from the above results, that the killed bacteria containing toxin provide many advantages, while still retaining all or a substantial proportion of the activity of the pesticidal activity of the live bacteria. Maintenance of dead bacteria for storage, transportation and application is more convenient and economic than employing the live organisms. At the same time, the organisms provide encapsulated protection of the toxins, so as to maintain the toxins' bioactivity for long periods of time. In addition, denaturation of the proteases prevents proteolytic degradation of the polypeptide pesticides. Various means for killing the organisms without disrupting their structure or denaturing the protein pesticide can be employed to provide the desired pesticidal activity, while providing for a convenient pesticidal composition for use in formulation and application to plants and plant products.

Although the foregoing invention has been described in some detail by way of illustration and example for purposes of clarity of understanding, it will be obvious that certain changes and modifications may be practiced within the scope of the appended claims.

We claim:

1. A pesticidal composition comprising pesticide-containing substantially intact microbial cells and an inert carrier wherein:

(1) the microbial cells have been stabilized with a chemical reagent selected from the group consisting of halogenating agents, aldehydes, anti-infective, alcohols and histological fixatives under conditions such that the cells have prolonged pesticidal activity when applied to the environment of a target pest; and

(2) the pesticide is a polypeptide, which remains intracellular after stabilisation, and is produced as a result of expression of a heterologous gene in the microbial cell, and which pesticide becomes accessible to a target pest upon ingestion of the cell by a pest.

2. A pesticidal composition, according to claim **1**, wherein said cells are prokaryotes or lower eukaryotes.

3. A pesticidal composition, according to claim **2**, wherein said prokaryote cells are selected from the group consisting of Enterobacteriaceae, Bacillaceae, Rhizobiaceae, Spirillaceae, Lactobacillaceae, Pseudomonadaceae, Azotobacteraceae, and Nitrobacteraceae.

4. A pesticidal composition, according to claim **2**, wherein said lower eukaryotic cells are selected from the group consisting of Phycomycetes, Ascomycetes, and Basidiomycetes.

5. A pesticidal composition, according to claim **1**, wherein said cells are a pigmented bacteria, yeasts, or fungi.

6. A pesticidal composition, according to claim **1**, wherein said microbial cells are Pseudomanas.

7. A pesticidal composition, according to claim **1**, wherein said substantially intact microbial cells are stabilized with formalin.

8. A pesticidal composition, according to claim **1**, wherein said substantially intact microbial cells are stabilized with iodine.

9. A pesticidal composition, according to claim **1**, wherein said microbial cells are is Pseudomonas and said toxin is a *B. thuringiensis* toxin.

10. A pesticidal composition, according to claim **1**, wherein said pesticide polypeptide is a *Bacillus thuringiensis* pesticide toxin.

11. A method of protecting a host susceptible to an invertebrate pest which comprises applying to the environment of the pest an effective amount of a pesticidal

composition which comprises pesticide-containing substantially intact microbial cells wherein:

(1) the microbial cells have been stabilized with a chemical reagent selected from the group consisting of halogenating agents, aldehydes, anti-infectives, alcohols and histological fixatives under conditions such that the cells have prolonged pesticidal activity when applied to the environment of a target pest; and

(2) the pesticide is a polypeptide, which remains intracellular after stabilisation, and is produced as a result of expression of a heterologous gene in the microbial cell, and which pesticide becomes accessible to the target pest upon ingestion of the cell by a pest.

12. A method, according to claim **11**, wherein said microbial cells are prokaryotes or lower eukaryotes.

13. A method according to claim **12**, wherein said prokaryotic cells are selected from the group consisting of Enterobacteriaceae, Bacillaceae, Rhizobiaceae, Spirillaceae, Lactobacillaceae, Pseudomonadaceae, Azotobacteraceae, and Nitrobacteraceae.

14. A method, according to claim **12**, wherein said lower eukaryote microorganisms are selected from the group consisting of Phycomycetes, Ascomycetes, and Basidiomycetes.

15. A method according to claim **11**, wherein said microbial cells are pigmented bacteria, yeasts, or fungi and said pesticide is an insecticidal polypeptide.

16. A method, according to claim **11**, wherein said microbial cells are stabilized by treatment with formalin.

17. A method, according to claim **11**, wherein said microbial cells are stabilized by treatment with iodine.

18. A method, according to claim **11**, wherein said host susceptible to an invertebrate pest is a plant.

19. A method, according to claim **11**, wherein said pesticide polypeptide is a *Bacillus thuringiensis* pesticide toxin.

* * * * *

United States Patent [19]

Cardina et al.

[11] **Patent Number: 4,643,756**

[45] **Date of Patent: Feb. 17, 1987**

[54] **BIOHERBICIDE FOR FLORIDA BEGGARWEED**

[75] Inventors: **John Cardina; Robert H. Littrell,** both of Tifton, Ga.; **Larry J. Stowell,** San Diego, Calif.

[73] Assignees: **Mycogen Corporation,** San Diego, Calif.; **Univ. of Georgia Research Foundation, Inc.,** Athens, Ga.; **The United States of America as represented by the Secretary of Agriculture,** Washington, D.C.

[21] Appl. No.: **715,229**

[22] Filed: **Mar. 22, 1985**

[51] **Int. Cl.**[4] **A01N 63/00**

[52] **U.S. Cl.** **71/79;** 435/911

[58] **Field of Search** 71/79; 435/911; 47/DIG. 11

[56] **References Cited**

U.S. PATENT DOCUMENTS

3,849,104	11/1974	Daniel et al.	71/79
3,999,973	12/1976	Templeton	71/79
4,162,912	7/1979	Charudatan	71/79
4,263,036	4/1981	Charudattan	71/79
4,390,360	6/1983	Walker	71/79
4,419,120	12/1983	Walker	71/79

OTHER PUBLICATIONS

Staples et al, "Appressorium Formation, etc.," (1976), Archives of Microbiology, vol. 109, pp. 75–84 (1976).

Almeida et al, "Curative and Preventative, etc.," (1981) CA 96: 29819s (1982).

Malhotra et al, "Utilization of Oligosaccharides, etc.," Pantnagar J. Res., (1977) 2(2), pp. 119–122.

Primary Examiner—Glennon H. Hollrah

Assistant Examiner—Robert Lelkes

Attorney, Agent, or Firm—Roman Saliwanchik; David R. Saliwanchik

[57] **ABSTRACT**

The subject invention concerns a novel bioherbicide and its use to control a major weed found in many fields in the Southeastern United States where peanuts and soybeans are grown. Specifically, *Colletotrichum truncatum* (Schw.) Andrus & Moore, in an agricultural composition, can be used to effectively control Florida beggarweed without adversely affecting field crops, e.g., peanuts and soybeans. Further, *C. truncatum* (Schw.) Andrus & Moore in a mixture with *Alternaria cassiae* can be used to control Florida beggarweed and other undesired vegetation, such as sicklepod, showy crotalaria and coffee senna.

6 Claims, No Drawings

(Continued on next page)

1

BIOHERBICIDE FOR FLORIDA BEGGARWEED

BACKGROUND OF THE INVENTION

Florida beggarweed (*Desmodium tortuosum* (SW.)DC) is a major weed problem in many fields in the Southeastern United States where peanuts and soybeans are grown. It is ranked as the most troublesome weed in peanuts in the three largest peanut producing states, and its spread is causing increasing concern to farmers throughout the Southeast. This weed occurs primarily in field crops, but is found also in fallowed areas and along roadsides and waste areas. Florida beggarweed is an annual herbaceous legume native to Florida and the Gulf States, extending to the subtropics and tropics of the Western Hemisphere. Although it is currently considered one of the most damaging weeds, particularly in leguminous crops, Florida beggarweed was a popular forage crop prior to 1950. As a forage crop, heavy seeding on land of moderate to high fertility was recommended so that when plants matured 70–80 days later, the crop could be expected to yield 4 to 6 tons of dry matter and return to the soil up to 5000 seeds per plant. As a result, Florida beggarweed, no longer considered an acceptable forage crop, plagues farmers throughout the Southeast on soils that are most suitable for peanuts and soybeans. The damage caused by Florida beggarweed is primarily a result of competition with the field crop for light, water, and nutrients. In addition, this weed, which may reach 10 feet at maturity, overtops the crop canopy and interferes with pesticidal spray programs. Particularly serious is the decreased coverage of biweekly foliar fungicide applications in peanuts caused by infestations of Florida beggarweed. In peanut production, where the crop must be inverted at the end of the season, Florida beggarweed interferes with the digging, turning and curing processes, and thereby causes further reductions in crop yield and quality.

Florida beggarweed harbors several pests of field crops. It is a common late-season food source for *Heliothis virescens* larvae, a pest of numerous crops. This weed is also a host of northern rootknot, lesion and sting nematodes that severely damage many crops. Recent evidence indicates that Florida beggarweed is a host of peanut mottle virus and peanut stripe virus.

Florida beggarweed is a leguminous weed that is not controlled by herbicide programs recommended for use in peanuts and soybeans. Development of chemical control for Florida beggarweed is complicated by its close botanical relationship to peanuts, soybeans, and other leguminous crops. This puts unusually stringent selectivity requirements on development of herbicides.

Chemical weed control programs are seriously inadequate for control of this weed. Florida beggarweed is resistant to many herbicides used in peanuts and soybeans. Frequently the weed germinates below the treated zone and avoids herbicide injury. Although many herbicides have been developed and tested in the last three decades, farmers still rely heavily on dinoseb, a herbicide developed in the 1950's. Dinoseb is a contact herbicide that causes injury to peanuts and soybeans; farmers tolerate this injury out of the necessity to control Florida beggarweed. Alternative approaches include MSMA (monosodium methanearsonate) paraquat, toxaphene and triazine herbicides, but these chemicals are not registered for use on these crops for reasons of toxicology and/or crop safety. Other inadequacies of chemical controls include lack of residual control, injury to nontarget organisms, undesirable residues in harvested products and carryover in subsequent crops.

2

The use of bioherbicides is becoming an increasingly important alternative to chemical herbicides for a variety of reasons, some of which are similar to those experienced in the chemical control of Florida beggarweed, as disclosed above. This importance is accompanied by several patents which have issued for bioherbicides and their use. Some of these patents, by way of illustration, are as follows: U.S. Pat. No. 3,849,104 (control of northern jointvetch with *Colletotrichum gloeosporioides* Penz. aeschynomene); U.S. Pat. No. 3,999,973 (control of prickly sida [teaweed] and other weeds with *Colletotrichum malvarum*); U.S. Pat. No. 4,162,912 (control of milkweed vine with *Araujia mosaic* virus); U.S. Pat. No. 4,263,036 (control of *Hydrilla verticillata* with *Fusarium roseum* Culmorum); U.S. Pat. No. 4,390,360 (control of sicklepod, showy crotalaria, and coffee senna with *Alternaria cassiae*); and U.S. Pat. No. 4,419,120 (control of prickly sida, velvetleaf, and spurred anoda with fungal pathogens.

BRIEF SUMMARY OF THE INVENTION

The subject invention concerns a novel bioherbicide which is effective, when used according to the process disclosed herein, in controlling Florida beggarweed without adversely affecting the growth and yield of the desired field crop. Preferably, the subject invention relates to the use of a composition comprising *Colletotrichum truncatum* (Schw.) Andrus & Moore spores in association with an agricultural carrier wherein said spores are in a concentration of from about 2×10^4 spores/ml of carrier to about 2×10^6 spores/ml of carrier, to control Florida beggarweed infestation. The concentrated spore formulation can be adapted for distribution over geographical locales or a situs where the spores germinate and infect Florida beggarweed.

DETAILED DESCRIPTION OF THE INVENTION

The fungus *Colletotrichum truncatum* (Schw.) Andrus & Moore is the first fungal pathogen that has been used to control Florida beggarweed. This fungus produces relatively insignificant leaf and stem lesions on mature plants, but causes a devastating disease on seedlings up to the 3–4 leaf stage. The initial symptoms of the disease are a twisting of stems and discoloration of leaf midrib and veins. Leaves eventually become mottled brown-yellow and die.

A subculture of *C. truncatum* (Schw.) Andrus & Moore has been deposited in the permanent collection of the Northern Regional Research Laboratory, U.S. Department of Agriculture, Peoria, Ill., USA, on Feb. 19, 1985. The culture was assigned the accession number NRRL 15933 by the repository. This deposit is available to the public upon the grant of a patent disclosing it. The deposit is also available as required by foreign patent laws in countries wherein counterparts of the subject application, or its progeny, are filed. However, it should be understood that the availability of a deposit does not constitute a license to practice the subject invention in derogation of patent rights granted by governmental action.

Taxonomic characteristics of *Colletotrichum truncatum* (Schw.) Andrus & Moore are as follows:

Colonies on V-8 and malt extract agars are composed of grayish-brown mycelium, becoming dark due to abundant formation of black sclerotia. The sclerotia are irregular in shape, composed of thick-walled, dark brown cells. Acervulate conidiomata erumpent through wheat stem epidermis contain a dense layer of conidiophores. Conidiophores are branched, consisting of dark brown hyphae bearing hyaline phialides at the apex. Numerous stiff, erect, tapered, dark brown setae are also produced in the conidiomata. Setae are variable in length, up to 235 μm long, with an average length of 165 μm. Conidia are hyaline, falcate, with narrowly rounded apices, guttulate, one-celled, but sometimes becoming septate prior to germination. The average size of 50 conidia was 30 μm long×4 μm wide (25–35×3.3–4.4 μm). Appresoria are dark, clavate to irregular in shape, averaging 9.2 μm long×7 μm wide (7.7–10.7×5.4–9.2 μm).

The conidial size is somewhat larger than that recorded by Sutton (Sutton, B. [1980] The Coelomycetes, Fungi imperfecti with pycnidia acervuli and stromata, Commonwealth Mycological Institute, Surrey, England) for this species, but it agrees better with the data from legumes reported by Tiffany and Gilman (Tiffany, L. H. and Gilman, J. C. [1954] Species of Colletotrichum from legumes, *Mycologia,* 46:52–75.

C. truncatum can be grown on solid or in liquid media. Solid media that can be used include water agar, potato dextrose agar, V-8 agar and string bean agar (strained extract of macerated string beans solidified in agar). Spores are produced on solid V-8 medium exposed to florescent light. Specifically, solid media can be, for example, (1) water agar, (2) potato dextrose agar (Difco), (3) lima bean agar (Difco), (4) corn meal agar (Difco), (5) potato-carrot agar (Tuite 19), and (6) Desmodium agar (blend 10 gm Desmodium plant parts or plant extracts in 1000 ml water and solidify with 20 gm agar).

For large scale production in fermentation tanks, liquid media is used, for example:

Formula I - Modified Richard's Solution - V-8*	
Sucrose	50 gm
Potassium nitrate	10 gm
Potassium phosphate, monobasic	5.0 gm
Magnesium sulfate.7 H_2O	2.50 gm
Ferric chloride	0.02 gm
V-8 juice	15 ml
Distilled water to make	1000 ml

*Trademark, The Campbell Soup Company for mixed vegetable juices.

Formula II—Modified Richard's Solution—Distillers Solubles—Same as Formula I but substitute 15 gm Distillers solubles for V-8 juice.
Formula III—Modified Richard's Solution—Brewers yeast—Same as Formula I above but substitute 15 gm brewers yeast for V-8 juice.
Formula IV—Modifed Richard's Solution—Torula Yeast—Same as Formula I above but substitute 16 gm torula yeast for V-8 juice.
Formula V—Oatmeal Solution—2%—20 gm oatmeal blended in 1000 ml distilled water.
Formula VI—Oatmeal Solution—4%+2% sugar—
40 gm oatmeal
20 gm sucrose
1000 ml distilled water
Formula VII—Beggarweed Broth—
20 gm beggarweed leaves and petioles
10 gm potassium nitrate
2.5 gm magnesium sulfate $7H_2O$

10 gm potassium phosphate dibasic
1000 ml distilled water
Adjust final pH to 6.8

The preparation of spores is commenced in preseed liter flasks containing about 300 ml of liquid medium which have been inoculated with spores. The medium is incubated for 1–3 days with agitation at a temperature of about 26° to about 30° C.

The preseed is then transferred aseptically to 20-liter seed tanks with additional sterile medium as described above. The tanks are provided with sterile air and agitation. The cycle is continued at a temperature of about 26° to about 30° C. for 1 to 3 days.

Larger fermentors (250 liter) are aseptically inoculated with the seed tanks (entire contents), described above. Additional sterile medium, as used above, is added and the pH is adjusted to about 6.0. The fermentor is supplied with sterile air and agitation, and is maintained at a temperature of about 26° to about 30° C. for from 1 to 3 days. The fermentor is then harvested by filtering the contents to remove insoluble solids and mycelia growth. The filtered beer is then centrifuged, the supernatant is discarded, and the remaining spore concentrate is resuspended in water and centrifuged again. The supernate is again discarded and the spore concentrate is collected, placed in plastic bags and stored in ice. The concentrate so stored maintains an 80% germination for up to 21 days.

The spore concentrate is mixed with an agriculturally acceptable diluent or carrier for application to the undesired host vegetation or a situs. By the term "situs" is meant those areas infested with the undesired vegetation or potential infestation sites.

The preferred carrier is water, and the spore concentrate is dispersed to make a concentration of from about 2×10^4 to 2×10^6 spores/ml. The formulation is then sprayed on the undesired vegetation or situs by conventional spraying equipment in an amount of from about 50 to 500 liters per hectare. Applications have been made to Florida beggarweed in the cotyledon to 6-leaf stage at temperatures from 20° to 30° C. Applications are most effective on seedlings with 1–3 leaves to 25° to 30° C.

The fungus is not pathogenic to "Florunner" peanuts or "Centennial" soybeans with spore concentrations up to 2×10^5.

Spores of *C. truncatum* (Schw.) Andrus & Moore can be mixed with those of *Alternaria cassiae* to enlarge the scope of control of undesired vegetation. For example, this mixture can be used to control both Florida beggarweed and sicklepod (*Cassia obtusifolia*), the two most troublesome weeds in soybeans and peanuts in the Southeast. Further, spores of *C. truncatum* (Schw.) Andrus & Moore can be mixed with those of *A. cassiae* to control Florida beggarweed adn showy crotalaria or coffee senna. The use of *A. cassiae* to control sicklepod, showy crotalaria and coffee senna is disclosed in U.S. Pat. No. 4,390,360, which is incorporated herein by reference thereto. The culture, means of growing, and application to these weeds disclosed in U.S. Pat. No. 4,390,360 can be used herein. Mixtures of *C. truncatum* and *A. cassiae,* for example, *A. cassiae* NRRL 12533, can be made by methods well known in the art, utilizing the disclosure of U.S. Pat. No. 4,390,360 and that contained herein.

Though spores are the preferred form of the fungi, the fungi also can be used in their vegetative form. For example, fragmented mycelia can be formulated and applied to Florida beggarweed and/or sicklepod, or showy crotalaria, or coffee senna in much the same manner as described above for the spore form.

Following are tabulated results showing the effectiveness of *C. truncatum* in controlling Florida beggarweed:

TABLE I

Effect of Inoculum Level on Anthracnose of Florida Beggarweed

Thousand spores per ml[1]	Proportion of plants dead in one week[2] %	Visual control rating[3] %	Plant survival after 12 days[4] %
0	0	0	100
25	49	62	35
50	57	65	30
100	57	74	30
250	60	81	20
500	74	97	6
1000	86	96	7

[1]Two-leaf stage plants sprayed to runoff 10-12-84, five replications.
[2]Based on actual counts.
[3]Visual estimate of % diseased tissue.
[4]Proportion of plants with at least one normal leaf.

TABLE II

Temperature C.	Dead %	Necrotic Tissue %
20	17.32	6.26
25	24.70	7.15
30	19.12	5.85

Data are based on plants grown in three growth chambers at the temperatures indicated. Plants were sprayed to runoff with 5×10^4 spores per ml, and covered with plastic bags for 48 hr. Ratings were made 7 days after treatment. Data are means of 3 replications.

TABLE III

Effect of Duration of Moist Period on Disease

Hours	Dead %	Necrotic Tissue %
4	0	0.275
8	0	1.988
12	3.365	4.859
24	24.401	9.174
48	45.068	12.159

Data are means of four replications of growth chamber-grown plants treated in the 3–4 leaf stage with 5×10^4

spores per ml to runoff and covered with plastic bags for the time indicated.

We claim:

1. A composition for agricultural application for controlling Florida beggarweed comprising an isolate of the fungus *Colletotrichum truncatum* (Schw.) Andrus & Moore having the identifying characteristics of culture deposit NRRL 15933, which isolate controls Florida beggarweed without adversely affecting peanuts or soybeans, in association with an inert agricultural carrier.

2. A composition, according to claim **1**, wherein said isolate is in the spore form at a spore concentration of from about 2×10^4 to about 2×10^6 spores/ml of carrier.

3. A process for controlling Florida beggarweed which comprises applying an effective amount of an isolate of the fungus *Colletotrichum truncatum* (Schw.) Andrus & Moore having the identifying characteristics of culture deposit NRRL 15933, which isolate controls Florida beggarweed without adversely affecting peanuts or soybeans, onto said Florida beggarweed or unto the situs of said Florida beggarweed.

4. A composition for agricultural application for controlling Florida beggarweed and sicklepod or showy crotalaria or coffee senna comprising a mixture of the fungi *Colletotrichum truncatum* (Schw.) Andrus & Moore and *Alternaria cassiae* having the identifying characteristics of culture deposit NRRL 12553 in association with an inert agricultural carrier.

5. A process for controlling Florida beggarweed and sicklepod, or showy crotalaria, or coffee senna which comprises applying an effective amount of a mixture of fungi consisting of an isolate of *Colletotrichum truncatum* (Schw.) Andrus & Moore having the identifying characteristics of culture deposit NRRL 15933, which isolate controls Florida beggarweed without adversely affecting peanuts or soybeans, and *Alternaria cassaie* having the identifying characteristics of culture deposit NRRL 12553 onto said Florida beggarweed and sicklepod, or showy crotalaria, or coffee senna or unto the situs of said Florida beggarweed and sicklepod, or showy crotalaria, or coffee senna.

6. A novel *Colletotrichum truncatum* characterized as having the property of controlling Florida beggarweed without adversely affecting peanuts and soybeans, and, when in its essentially biologically pure form, has the taxonomic characterization of *C. truncatum* (Schw.) Andrus & Moore culture deposit NRRL 15933.

* * * * *

United States Patent [19]

Anilionis et al.

[11] **Patent Number: 4,721,671**

[45] **Date of Patent: Jan. 26, 1988**

[54] EFFICIENT PROKARYOTIC EXPRESSION SYSTEM USING PORTIONS OF THE *E. COLI* βGLUCURONIDASE GENE

[75] Inventors: **Algis Anilionis; John L. Palmer,** both of Arlington, Mass.

[73] Assignee: **Repligen Corporation,** Cambridge, Mass.

[21] Appl. No.: **686,344**

[22] Filed: **Dec. 26, 1984**

[51] **Int. Cl.**[4] **C12P 21/00;** C12N 15/00; C12N 9/26; C12N 1/20; C12N 7/00; C07H 15/12; C07K 13/00

[52] **U.S. Cl.** **435/68;** 435/172.3; 435/201; 435/253; 435/320; 435/849; 536/27; 935/14; 530/350

[58] **Field of Search** 435/70, 172.3, 317, 435/201, 253; 935/14; 536/27

[56] **References Cited**

PUBLICATIONS

Hieber, Biochem. and Biophys. Res. Comm., 104(4):1271–1278, 1982, (Feb.).

Ashwell, G. (1962), Methods in Enzymol., 5:190–208.

Ritzenthaler, P. et al., (1983), Mol. Gen. Genet., 191:263–270.

Novel, M. and Novel, G., (1976), J. Bacteriol., 127:406–432.

Blanco, C. et al., (1982), J. Bacteriol., 149:587–594.

Clarke, L. and Carbon, J., (1976), Cell, 9:91–99.

Primary Examiner—Thomas G. Wiseman
Assistant Examiner—Thomas D. Mays
Attorney, Agent, or Firm—Roman Saliwanchik; David R. Saliwanchik

[57] **ABSTRACT**

A novel highly effective prokaryotic expression system is exemplified specifically by being used to produce the useful enzyme β-glucuronidase (BG). This system uses a hybrid plasmid comprising BG gene promotor DNA. The level of expression of BG by an *E. coli* K-12 derivative host is in the 50% of total cellular protein range. The invention expression system also can be used to express other useful proteins, as disclosed herein.

27 Claims, 1 Drawing Figure

Figure 1

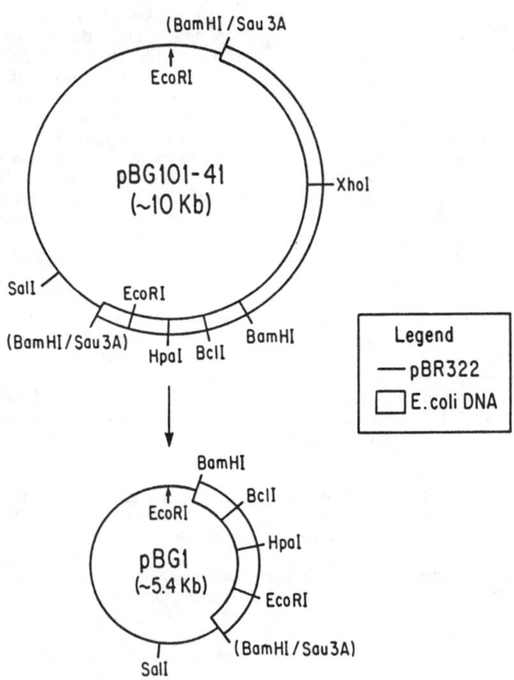

1

EFFICIENT PROKARYOTIC EXPRESSION SYSTEM USING PORTIONS OF THE *E. COLI* βGLUCURONIDASE GENE

BACKGROUND OF THE INVENTION

Expression of cloned DNA in bacteria has been extensively studied. Most available information concerns expression of cloned native, foreign, and hybrid genes in *E. coli* using portable promoters in plasmids, phage or cosmid cloning vectors. Recently several laboratories have reported expression levels of 30% of total cellular protein as a single protein product.

The *E. coli* gene uidA encodes the enzyme β-D-glucuronidase (E.C. 3.2.1.31) which is the first enzyme of the hexuronide-hexuronate pathway (Ashwell, G. [1962] Methods in Enzymol. 5:190–208). The enzyme is induced by β-D-glucuronides but hydrolyzes both β-D-glucuronides and the non-inducing β-D-galacturonides to their respective uronic acids. Subsequent enzymes of the pathway convert glucuronate or galacturonate to 2-keto-3-deoxy-gluconate (KDG), which is in turn fed into the glycolytic pathway after phosphorylation to produce KDG-6-P and conversion to pyruvate plus glyceraldehyde-3-phosphate.

The uidA gene has been shown to be under negative regulation by the closely linked upstream repressor gene uidR and also under partial or weaker negative control by the uxuR gene (Ritzenthaler, P., Blanco, C. and Mata-Gilsinger, M. [1983] Mol. Gen. Genet. 191:263–270). uxuR is the repressor of the uxu operon which produces the enzymes necessary for the conversion of fructuronate to KGD later in the same pathway (Novel, M. and Novel, G. [1976] J. Bacteriol. 127: 407–417; ibid, 418–432).

Blanco et al. (1982, J. Bact. 149: 587–594) identified a plasmid U1, from the *E. coli*-ColE1 hybrid clone bank of Clarke and Carbon (1976, Cell 9: 91–99), which carries the manA, uidA, uidR region of the *E. coli* chromosome. When fully induced, strain JA200(U1) produced only 1.8-fold the β-D-glucuronidase enzyme induced in JA200. In contrast, subclones which carried the uidA gene without a uidR repressor gene copy produced sufficient enzyme to be visualized by Coomassie blue staining of a non-denaturing polyacrylamide gel (FIG. 3 in Blanco et al.).

BRIEF SUMMARY OF THE INVENTION

The subject invention concerns a newly recognized, highly efficient expression system which enables workers skilled in the art to achieve heretofore unattained levels of expression of cloned gene products, i.e., over 50% of total cellular protein of the bacterial host comprising a single polypeptide product. The expression system is exemplified herein to produce high yields of the useful enzyme β-glucuronidase (BG). This expression of high amounts of BG is achieved by use of a hybrid plasmid vector comprising *E. coli* BG gene DNA. It is surprising that the *E. coli* host transformed with the novel hybrid plasmid of the subject invention expresses BG in extremely high amounts in view of the known facts that BG is expressed in minute amounts by its native *E. coli* host, and that the known *E. coli* clones of the BG gene described by Blanco et al. (1982) do not express the BG enzyme at these ultra-high levels. It is believed that this prior art level of expression of BG by *E. coli* has led persons skilled in the art away from using BG promoter DNA in prokaryotic expression systems.

2

Rather, the well-characterized lac and trp promoters have been extensively used in prokaryotic expression systems.

Expression vectors which exemplify the subject high expression system invention are hydrid plasmids pBG101-41 and pBG1. Novel hybrid plasmid pBG101-41 contains approximately 6 kb of *E. coli* BG DNA inserted at the BamHI site of pBR322. Novel hybrid plasmid pBG1 contains the 1237 base pair SalI-BamHI fragment of pBG101-41 ligated into a SalI, Bam HI cut pBR322 vector. Plasmid pBG1 carries all the nucleotide bases necessary to achieve the high level expression of a host transformed with plasmid pBG101-41. See Chart A.

DESCRIPTION OF THE DRAWING AND CHART

The Drawing depicts the endonuclease restriction map of plasmids pBG101-41 and pBG1.

The Chart depicts the *E. coli* BG insert DNA sequence of pBG1. The 26 bp dyad is outlined by dashed lines. RBS = ribosome binding site.

DETAILED DISCLOSURE OF THE INVENTION

The subject invention, which is in the field of molecular biology, concerns expression vectors useful to produce the useful enzyme β-D-glucuronidase in a highly efficient manner. The level of expression of β-D-glucuronidase by a suitable *E. coli* host is at least 50% of total cellular protein.

β-D-glucuronidase is a useful biological tool. For example, it can be used in the determination of urinary steroids and of steroid conjugates in blood. The current high cost of this enzyme is due to its minute presence in native *E. coli*. This invention greatly decreases the cost of production of this enzyme as the enzyme is produced at a level of 50% of total protein in the cell culture. This increase in expression level impacts cost of protein production in two independent and significant ways. Increased expression greatly increases the amount of the useful enzyme that can be produced in a growth vessel, for example a fermenter, of any given size. In addition, an increase in expression level greatly decreases the degree of difficulty encountered in the purification of the enzyme. In many of the applications of this useful protein it must be in pure form.

The expression system of the subject invention can be used to express other useful proteins as well as β-glucuronidase. These proteins can be expressed at high level either by inserting their gene behind the BG promoter or by making a hybrid gene which contains the BG promoter and the N-terminal sequences of the BG gene fused to coding sequences of the other gene. Useful proteins which can be expressed at elevated levels by one or both of these methods include industrially useful enzymes, for example, sugar isomerase (glucose isomerase); proteases; sugar and starch amylases; rennet; esterases; and oxygenases, e.g., lignin degrading enzymes and aromatic ring-cleaving enzymes. Other useful proteins which may be expressed with this system include hormones, e.g., insulin, interferons, interleukins, growth hormones, including human, bovine and chicken, and somatomedins; platelet derived growth factor; tumor necrosis factor; glucagon; hypothalamic hormones (growth hormone stimulating factor, ACTH, endorphin); adrenal derived protein hormones; and rennin. In addition, this invention is of particular use in

3

the expression of bacterial toxins and of viral proteins. These viral proteins can be used in the production of vaccines and the development of assay kits that detect the presence of either the virus or a virus-derived antibody in a biological fluid. The bacterial toxin expressed by this system can be used for the treatment of disease or for the control of pests in the environment. All of the proteins discussed above can be produced by the techniques discussed above and by other techniques that are known to persons skilled in the art.

Before detailing the construction and identity of the novel plasmids and expression system of the subject invention, there are disclosed the materials and methods employed.

(1) MEDIA

Cultures were grown in YT medium (8 gm tryptone, 5 gm yeast extract, 5 gm NaCl per liter). Ampicillin was added to 50 μg/ml and tetracycline to 20 μg/ml as required. Glucuronidase screening plates contained 5 μg/ml μ-methyl-umbelliferyl-glucuronide (MUG) as described by Dahlen and Linde ([1973] Appl. Microbiol. 26: 863–866).

(2) *E. coli* DNA

E. coli DNA was prepared from strain MS371 essentially according to Marmur (1961, J. Mol. Biol. 3: 208–218) with modifications of Anilionis and Riley (1980, J. Bacteriol. 143: 355–365.) All *E. coli* strains disclosed herein are *E. coli* K-12 derivatives.

(3) β-D-GLUCURONIDASE ASSAYS

Hydrolysis of p-nitrophenol-62-glucuronide (pNPGA) and liberation of free p-nitrophenol was measured by absorbance increase at 400 nm. Assays were calibrated with standard *E. coli* type VII enzyme stocks obtained from Sigma Chemical (St. Louis, MO). The Sigma unit definition of β-D-glucuronidase (the modified Fishman unit) is the activity that will liberate 1.0 μg phenolphthalein from phenolphthalein glucuronide in one hour at 37° C. at the optimum pH (6.8 for the *E. coli* enzyme). The assay procedure was as follows: Each assay tube contained 0.98 ml 50 mM NaH_2PO_4 (pH 6.8) pre-equilibrated with the enzyme sample or 10 units of *E. coli* type VII enzyme standard in 0.01 ml for 10 min at 37° C. The reactions were started by adding 0.1 ml 10 mg/ml pNPGA substrate solution, quickly mixed, and incubated at 37° C. in a water bath. Reactions were stopped by the addition of 0.1 ml 1.0N NaOH. The absorbance at 400 nm was measured against a blank containing all components except enzyme incubated for the same period at 37° C. Duplicate assays were performed for each of three different times to obtain accurate activity data.

(4) TWO-DIMENSIONAL GEL ELECTROPHORESIS

Lysates of *E. coli* were prepared by treatment with lysozyme followed by several freeze-thaws. Clarified supernates of these lysates were assayed for protein. These samples were then brought to 2% NP40 and 5% β-mercaptoethanol and loaded onto an isoelectric focusing gel.

The isoelectric gels were cast in acid-washed 160×2.5 mm (ID) glass tubes. Formulation of the gel solution was as follows (for 10 ml):

5.5 g urea (ultrapure, Schwarz/Mann, Spring Valley, N.Y.)

4

1.33 ml acrylamide solution (28.4% w/v acrylamide 1.6% w/v methylene bisacrylamide

2.0 ml 10% v/v NP40

1.97 ml H_2O

0.60 ml ampholytes (LKB, Gaithersburg, MD), pH 3–10

0.008 ml tetraethyl methylene diamine

Following degassing of the solution, 10 μl of a 10% w/v solution of ammonium persulfate was added and approximately 1.2 ml was added to each glass tube via a 4 in or greater spinal-tap needle. The bottoms of the tubes were previously sealed with parafilm. The tope of the gel was overlaid with 20 μl of deionized water and polymerization occurred within 30 min at room temperature. The isoelectric focusing gels were placed in a tube gel electrophoresis apparatus (Hoefer Scientific Instruments, San Francisco, CA). The lower (anodic) electrode buffer was 10 mM phosphoric acid and the upper (cathodic) electrode solution was 20 mM NaOH. The NaOH solution was freshly made or stored under vacuum to prevent CO_2 buildup. The gels were prefocused at 300 v for 30 min prior to sample loading.

The samples (usually 100 μg protein) were layered onto the top (cathodic end) of the isoelectric focusing (IEF) gel. IEF was then carried out for 14 hr at 300 v and 2–3 hr at 800 v. The gels were extruded from the tubes and equilibrated for 30 min with a solution comprising 2.3% SDS, 62.5 mM Tris-HCl (pH 6.8), and 5% β-mercaptoethanol. The equilibrated gels were laid on top of the stacking gel of a conventional discontinuous SDS-polyacrylamide slab gel and sealed in place with a hot 1% agarose solution made up in equilibration buffer (2.3% SDS, 62.5 mM Tris-HCl (pH 6.8), 5% β-mercaptoethanol) also containing 0.01% bromophenol blue as a tracking dye.

The gels were run in the second dimension at 25 mamp gel until the tracking dye reached the end of the gel. The slab gels were then stained with 0.24% Coomassie blue R in 50% methanol, 10% acetic acid for about 30 min at room temperature with gentle agitation. Destaining was achieved overnight in 10% methanol, 5% acetic acid, also with gentle agitation, and stained gels were then photographed directly.

(5) PLASMID DNA PREPARATION

Procedure used for large scale preparation of plasmid DNA was essentially as follows: A 250 ml culture was grown to Log phase, amplified with chloramphenicol at O.D. 0.6 to 0.7 (or alternatively with no chloramphenicol addition) and grown overnight. Cells were pelleted at 6K, 20 min, JA14 rotor, and resuspended in 6 ml glucose buffer (50 mM glucose, 25 mM tris, 10 mM EDTA). Cells were incubated 10 min at room temp in the presence of 1 ml of 20 mg/ml lysozyme freshly made; placed on ice with the addition of 13.8 ml 1% SDS in 0.2N NaOH for 5 min, and kept on ice an additional 15 min with 7 ml 5M KAC (pH 5.0–5.5). Debris was pelleted at 10K for 10 min and supernate extracted once with an equal volume of pheno-chloroform-isoamyl alcohol (25:24:1, TE saturated, 0.1% 8-hydroxyquinoline). Following precipitation with 0.6 vol. isopropyl alcohol, DNA was purified over CsCl gradients.

(6) RESTRICTION ENZYME DIGESTION AND ISOLATION OF DESIRED FRAGMENTS

Digestions were carried out according to supplier's instructions. Separation of fragments was achieved by agarose gel electrophoresis in TBE buffer (90 mM Tris,

0.89M borate, 2 mM EDTA) containing 0.5 μg ethidium bromide/ml. Isolation of the desired fragment was achieved by excision of the appropriate region of the gel and subsequent electroelution of the DNA in 1X TBE, at 100 volts for 2 hr followed by a reversal of current for 1 min to decrease adherence of the DNA to the dialysis tubing. Eluted DNA was purified and concentrated by passing over Elu-tip columns (Schleicher and Schuell, Keene, NH) according to supplier's instructions, followed by precipitation in 2.5 volumes EtOH with added carrier tRNA.

(7) DNA LIGATIONS

T4 ligase was used for standard vector/insert ligations and was present in excess (200 u/μg DNA); insert was present in a 5-fold molar excess with 0.02 pmoles of vector and 0.1 pmoles of insert in a 20 μl reaction volume.

(8) TRANSFORMATION

Fresh overnight cultures were diluted in L-broth and allowed to grow at 37° C. with agitation until an A_{600} of 0.3 was obtained. The cells were chilled on ice, then collected by centrifugation (10 min at 4100 xg). The cells were resuspended in ½ the original volume of ice cold 50 mM $CaCl_2$ and incubated on ice for 20 min. The cells were again collected by centrifugation as above and resuspended in ice cold 50 mM $CaCl_2$ (1/25 the original volume). 0.1 ml of the cell suspension was mixed with 1–10 μl (50–100 ng) of DNA plasmid solution and allowed to sit for 30 min at 0° C. The cells were then heated to 37° C. for 2 min and plated on L-broth plates containing 1.5% agar and either 10 μg/ml tetracycline or 50 μg/ml ampicillin. The plates were incubated overnight at 37° C. Transformation efficiencies of 1×10^6 colonies per μg plasmid DNA were routinely observed.

(9) AGAROSE ELECTROPHORESIS

DNA fragments were isolated by gel electrophoresis in 0.8% agarose in 2X tris-borate buffer (178 mM tris, 178 mM boric acid, 5 mM Na_2EDTA pH 8.4). Analytical and preparative gels were run in a horizontal gel box at 60 volts submerged in electrophoresis buffer (1X tris-borate). DNA bands were visualized under UV light by including 5.0 μg/ml ethidium bromide (EtBr) in the gel. A slice containing the desired DNA band was cut from the gel and the DNA recovered by electrophoresis in 1X tris-borate buffer in a dialysis tube (½ in. diameter) containing 0.5–1.0 ml of buffer. Electrophoresis was carried out for 30 min at 10 volts or until the stained material was located against the side of the dialysis tubing. The gel slice was removed from the dialysis bag and the DNA recovered by repeatedly flushing the bag with tris-borate buffer. NaCl was added to the DNA solution to a final concentration of 1M and the ethidium bromide and agarose gel impurities were removed by two extractions with phenol saturated with tris borate buffer. The phenol was removed by two extractions with ether and the purified DNA was recovered by precipitation with 1/10 volume 3M sodium acetate, pH 4.5 and 2.5 volumes cold ethanol. The precipitation reaction was carried out at −70° C. for 15–20 min. The precipitated DNA was recovered by centrifugation at 10,000 xg for 15 min. Yield of recovered fragment was assayed by direct comparison of ethidium bromide fluorescence with pure DNA standards. Typi-

cally, 50% recoveries were obtained with the yield decreasing as fragment size increased.

(10) POLYACRYLAMIDE GEL ELECTROPHORESIS

All SDS gels were run by the method of Laemmli (Laemmli, U.K. [1970] Nature [London] 227: 680–685). These gels contained a total acrylamide concentration of 9%. Slab gels were 1.5 mm wide, run in an electrophoretic apparatus obtained from Hoefer Scientific Instruments (San Francisco, Calif.). Tube gels were run in 6 mm i.d. × 10 cm glass tubes without a stacking gel.

Tube gels were stained by the following method: Gels were stirred overnight in 650 ml isopropyl alcohol, 250 ml H_2O, and 100 ml acetic acid, containing 1.0 g/l Coomassie blue. Then they were stirred overnight again in 100 ml acetic acid, 100 ml isopropyl alcohol, and 800 ml H_2O, containing 0.05 Coomassie blue. A third overnight stirring was done in 10% acetic acid. The gels were then scanned using a Beckman model 34 spectrophotometer equipped with a model 576767 gel scanner (Beckman Instruments). Percent expression was determined by cutting out the protein peaks and weighing the paper. Percent hybrid protein expression is calculated as [weight of β-glucuronidase protein/weight of total protein peaks] × 100%.

(11) BACTERIAL STRAINS AND MEDIA

The source and genotype of all bacterial strains used are listed infra. All strains were maintained and grown using YT medium (8 gm/l tryptone, 5 gm/l yeast extract, and 5 gm/l sodium chloride).

(12) CHEMICALS

Growth media components were obtained from Difco (Detriot, MI). Acrylamide was obtained from Accurate Chemical & Schientific Corp., (Westbury, N.Y.). All other chemicals were obtained from Sigma Chemical Co. (St. Louis, Mo.).

(13) CULTURES (ALL *E. COLI* K-12 STRAINS)

(A) Bacterial Strains	Relevant Genotype	Repository Number
E. coli MS371	F^-, Gal^-, Thi^-, endA, sbcB, hsdR4	NRRL B-15129 Deposited Aug. 18, 1982 and now available to the public.
E. coli GMS407	argE3, lacY1, galK2, man4, uidA1, mtl1	NRRL B-15917 Deposited on Dec. 7, 1984.
E. coli HB101	Pro^-, Leu^-, Thi^-, lacY, hsdR, endA, recA, rpsL20, supE44	NRRL B-11371 Available to the public

(B) Bacterial Host Containing Plasmid	Repository Number
MS371(pBG101-41)	NRRL B-15905 Deposited Nov. 1, 1984
MS371(pBG1)	NRRL B-15904 Deposited Nov. 1, 1984

(C) Plasmid

Plasmid pBR322 is a well-known and available plasmid. It is maintained in the *E. coli* host ATCC 37017. Purified pBR322 DNA can be obtained as described in Bolivar, F., Rodriquez, R. L., Greene, P. J., Betlach, M. C., Heyneker, H. L., Boyer, H. W., Crosa, J. H. and Falkow, S. (1977) Gene 2:95–113; and Sutcliffe, J. G. (1978) Nucleic Acids Res. 5: 2721–2728.

NRRL B-15905, NRRL B-15904, and NRRL B-15917 are available to the public upon the grant of a patent which discloses these accession numbers. It should be understood that the availability of these deposits does not constitute a license to practice the subject invention in derogation of patent rights granted for the subject invention by governmental action. The culture deposits are in the permanent collection of the Northern Regional Research Laboratory (NRRL), U.S. Department of Agriculture, Peoria, Ill., U.S.A.

There are other well-known *E. coli* hosts which can be used instead of *E. coli* MS371, for example, *E. coli* RR1, HB101, and *E. coli* GMS407 (Novel, M. and Novel, G. [1973] Mol. Gen. Genet. 120: 319).

Further, other prokaryotic hosts which can be used are microbes from the genera Salmonella, Pseudomonas, Bacillus, Streptomyces, and the like.

(14) ISOLATION OF RECOMBINANT PLASMID DNA FROM TRANSFORMED HOST

Recombinant plasmid DNA can be isolated from its prokaryotic host by well-known procedures, e.g., using cleared lysate-isopycnic density gradient procedures, and the like.

(15) DNA SEQUENCING

DNA sequence determination was carried out as described by Maxam and Gilbert (1977, Proc. Natl. Acad. Sci. U.S.A. 74: 560–564) and Sanger, F. Nicklen, S. and Coulson, A. R. (1977) Proc. Natl. Acad. Sci. U.S.A. 74: 5463–5467.

It is within the skill of those in the art to vary the conditions required to grow cells, extract DNA, perform restriction enzyme digestions, electrophorese DNA fragments, tail and anneal plasmid and insert DNA, ligate DNA, transform cells, prepare plasmid DNA, electrophorese proteins, and sequence DNA.

(16) CONSTRUCTION OF HYBRID PLASMIDS

The source for the β-glucuronidase (BG) gene DNA used to construct hybrid plasmid pBG101-41 was *E. coli* MS371. This BG gene DNA was inserted into the BamH1 site of pBR322. Ligation mixtures were transformed into competent cells of *E. coli* which lack BG activity. Subsequent plating on YT agar, followed by screening and purification yield a clone designated BG101-41. Plasmid DNA was isolated from this clone and retransformed. The identity of plasmid pBG101-41 was then established by size of DNA, restriction endonuclease pattern and expression of BG in prokaryotic host.

Novel hybrid plasmid pBG1 was constructed from plasmid pBG101-41 by taking the 1237 bp SalI-BamHI fragment of pBG101-41 and ligating it into a SalI, BamHI cut pBR322 vector. The identity of pBG1 was established as above for pBG101-41.

Following are examples which illustrate procedures, including the best mode, for practicing the invention. These examples should not be construed as limiting. All percentages are by weight and all solvent mixture proportions are by volume unless otherwise noted.

EXAMPLE 1—CONSTRUCTION OF HYBRID PLASMID pBG101-41

Twenty μg of *E. coli* MS371 DNA was incubated at 37° C. with 2 units Sau3A enzyme in a reaction volume of 200 μl; 50 μl aliquots were removed and the reaction terminated by heating to 80° C. after 2 min, 5 min, 10

min, and 30 min. Examination of 1 μg samples after gel electrophoresis through 1% agarose showed that the 2 min and 5 min samples contained partial digests with average molecular lengths in the 5–10 kb range. One microgram amounts of each of these partial digests were ligated with 0.2 μg BamHI cleaved pBR322 which had also been heat-treated to inactivate the restriction enzyme. Ligation mixtures were transformed into competent cells of the *E. coli* mutant strain GMS407 (NRRL B-15917) which lacks β-glucuronidase activity and each mixture spread over 10 YT agar plates supplemented with 50 μg/ml ampicillin and 5 μg/ml MUG. After overnight incubation at 37° C., the plates were examined under ultraviolet light. Each plate was densely covered with colonies but on three plates a single focus of intense fluorescence was seen. Colonies from these foci were purified by restreaking on YT, ampicillin, MUG agar plates. After four rounds of purification yielding isolated intensely fluorescent colonies, one such purified clone, designated BG101-41, was chosen for further study. Isolation of plasmid DNA from this transformation and retransformation into *E. coli* strains GMS407, MS371, and HB101 showed that the MUG+ fluorescence phenotype was coselected with ampicillin resistance conferred by the plasmid pBG101-41.

A restriction site map of pBG101-41 plasmid DNA was derived and is presented in the drawing. The *E. coli* DNA insert was determined to be approximately 6 kb in extent and contain single recognition sites for the endonucleases BamHI, EcoRI, and XhoI. The insert was not cut by PstI, ClaI, HindIII, or SalI.

The presence of high levels of β-glucuronidase activity in pBG101-41 transformed strains was confirmed by the pNPGA assay. Lysozyme-EDTA lysed GMS407(pBG101-41) and MS371 (pBG101-41) cells showed similar activities of 1,000–1,500 units β-D-glucuronidase activity/mg wet weight cells. 100 μg protein samples of clarified lysates of YT grown stationary phase cultures of MS371 and MS371 (pBG101-41) were resolved by two-dimensional electrophoresis as described previously. The pattern of proteins seen in the MS371 lysate is that of a typical *E. coli* strain, whereas the MS371 (pBG101-41) lysate exhibits a single major protein species with a much diminished content of the normal complement of *E. coli* proteins. This major protein has an apparent molecular weight of 72Kd and an isoelectric point of approximately 6.8, consistent with the properties of *E. coli* β-glucuronidase enzyme. The major protein is clearly more abundant than any other protein in the cell and visual estimates suggest that it may in fact constitute over 50% of the total cellular protein. This corresponds to a level of over 6% of the wet weight of the cells.

Glucuronidase enzymes from other sources exist in tetrameric form leading to the prediction that the native *E. coli* enzyme should have a very high molecular weight and might be separated from other host proteins by gel filtration. Nine ml of crude lysate of MS371 (pBG101-41) cells containing 3.5×10^6 units of β-D-glucuronidase activity were mixed with 1.5 ml 0.5M sodium phosphate pH 6.8 and loaded onto an Ultragel TM ACA34 (LKB, Gaithersburg, Md.) column (2.5 cm×120 cm, 590 ml capacity) at a flowrate of 1 ml/min. The column was run in 50 mM sodium phosphate pH 6.8, 1 mM EDTA, 1 mM DTT. A single peak of partially included protein was β-glucuronidase activity eluted from the column well before the low molecular weight protein peak. The β-glucuronidase peak recovered showed a total activity of 2.1×10^7 units (a six-fold increase over the total apparent activity of the column load) and a protein content of 490 mg. The increase in apparent activity may be due to removal of inhibitory agents in the crude lysate from the enzyme preparation. The specific activity of this fraction was therefore 4.2×10^7 u/gm. Samples (20 μg) of four such purified enzyme preparations run on a 12% polyacrylamide gel were judged to be greater than 95% pure β-D-glucuronidase after only one column purification step.

EXAMPLE 2—CONSTRUCTION OF HYBRID PLASMID pBG1

Hybrid plasmid pBG1 was constructed as shown in the drawing. pBG1 is a clone of the 1237 bp SalI-BamHI fragment of pBG101-41 ligated into a SalI, BamHI cut pBR322 vector. This plasmid was transformed into MS371 and GMS407 and the total protein content of overnight cultures of these transformants were examined on SDS-polyacrylamide gels. Four isolates of MS371(pBG1) showed the presence of an abundant protein of approximately 20 Kd molecular weight which is not found in the MS371 parent. This protein is presumed to be the truncated uidA enzyme terminated after the BamHI site in pBG1, and accumulates to an estimated 15% of the total cellular protein, producing a 20 Kd truncated N-terminal fragment of the enzyme which has no β-glucuronidase activity. This corresponds to a similar number of molecules expressed per cell as for the clone BG101-41, since the truncated polypeptide is of the order of one quarter the size of the intact enzyme. The plasmid pBG1 thus carries all the sequences necessary to specify the high level expression

ies carry uid operator sites which bind the repressor molecules releasing repression of the host uidA gene.

Chart A shows the *E. coli* BG insert DNA sequence of BG1. The single open reading frame truncated at the BamHI site is the uidA structural gene. The nucleic acid sequence of the upstream control region is disclosed, along with the adjoining 170 codons of the N-terminal segment of the structural gene. The 17 N-terminal amino acid residues of the β-D-glucuronidase enzyme have been confirmed by amino acid sequencing of the isolated protein product.

Computer-aided searches of the DNA sequence of the uidA gene upstream control region reveal a potential promoter configuration, a presumptive CAP-binding site, a 26 bp imperfect dyad symmetry which may be involved in repressor binding, and a strong ribosome binding site with an appropriately spaced methionyl initiator codon.

Since plasmid pBG1 carries all the sequences necessary for high expression, it can be used as a transfer vector for the expression of useful proteins, as discussed previously. The procedures for use of the nucleotide sequence of pBG1 shown in Chart A to function as a promoter sequence in a variety of transfer vectors and prokaryotic hosts are well known and standard in the art. For example, the promoter DNA sequence of Chart A can be used in much the same manner as the lac promoter is presently used in the art. Further, it has been determined that only a portion of the nucleotide sequence shown in Chart A can be used as a promoter to produce a very high level expression (over 50% of total cellular protein) of a useful protein (protein A) in a prokaryotic (*E. coli*) expression system. We have shown that the following sequence confers extremely high expression activity:

```
NdeI                                         TaqI
CATATGTCAT  GAGAGTTTAT  CGTTCCCAAT  ACGCTCGAAC

GAACGTTCGG  TTGCTTATTT  TATGGCTTCT  GTCAACGCTG

 AhaIII               Sau3A
TTTTAAAGAT  TAATGCGATC  TATATCACGC  TGTGGGTATT

                      Sau3A
GCAGTTTTTG  GTTTTTTGAT  CGCGGTGTCA  GTTCTTTTTA

                  NcoI
TTTCCATTTC  TCTTCCATGG  GTTTCTCACA  GATAACTGTG

                     HpaI
TCCAACACAG  AATTGGTTAA  CTAATCAGAT  TAAAGGTTGA

                        RBS
CCAGTATTAT  TATCTTAATG  AGGAGTCCCT  T  ATG  TTA  CGT

                                                       TaqI
CCT  GTA  GAA  ACC  CCA  ACC  CGT  GAA  ATC  AAA  AAA  CTC  GAC  GGC
```

observed in clone BG101-41.

In order to check for the presence of an uidR operator site in pBG1 strains, GMS407 (pBG1) and MS371(pBG1) were examined on plates containing β-methylumbelliferyl-glucuronide (MUG). Fluorogenic activity was present in the MS371 strain but not in GMS407. Since the activity is not seen in the GMS407 host, it cannot be directly contributed by the plasmid, and since the activity appears in MS371(pBG1) the phenomenon must be due to derepression of the chromosomal uidA gene copy. Thus the plasmid pBG1 cop-

The above nucleotide sequence can be readily engineered by a person skilled in the art so that any part of the sequence can be used in an expression system to enhance the production of useful proteins. For example, the sequence can be cut at any one of the restriction sites shown, and at others not shown, to give a variety of segments of the sequence. Further, once the sequence is cut with one or more restriction endonucleases, the resulting fragments can be "chewed back" further by use of exonucleases; or synthetic linkers can be added. In addition a double-stranded synthetic oligonucleotide fragment incorporating a multiple cloning site comprising several restriction endonuclease recognition sequences could be inserted into the translated region and the cloned segment would then have a more general utility. These operations all can be done by a person skilled in the art without undue experimentation. Thus, the scope of the subject pioneer invention encompases all or part of the nucleotide sequence shown above, provided that which is used enhances the level of expression in the prokaryotic expression system employed. Again, whether there is an enhancement of the expression level in a prokaryotic expression system can be readily ascertained by a person skilled in the art, using standard procedures described herein, without undue experimentation.

It should also be clear that the scope of the subject invention includes use of all or part of the disclosed BG structural gene sequence and all or part of the upstream untranslated DNA sequence of the BG promoter region not only to produce BG but also as part of an expression system to produce other useful proteins, many of which are exemplified herein. In using the subject invention thusly, a person skilled in the art could readily determine whether all or only a part of the BG promoter and gene DNA would function best to give the high expression levels of desired protein. This determination, which can be done without undue experimentation, may well show that part or all of the 271 base pairs of the upstream untranslated sequence disclosed above, along with all or part of the first 17 codons of the BG structural gene shown in Chart A and reproduced above, would give the highest expression level in a particular expression system designed for the production of a specific useful protein.

It should be readily evident to a person skilled in the art that the proteins produced by the BG expression system of the subject invention may be hybrid proteins, in that DNA from the expression system would be transcribed along with the particular protein gene DNA. These hybrid proteins would still retain the primary function(s) of the desired protein and, thus, could be used as the desired protein is used. If preferred, it is well within the skill of those in the art to further process the hybrid protein and recover the desired protein.

As is well known in the art, the nucleic acid sequence upstream of the structural gene specifies its promoter. Only certain of these nucleotides are important for promoter activity: those located 35 bases upstream, and those located 10 bases upstream of the site at which the mRNA is initiated. See, for example, Hawley, D. K. and McClure, W. R. (1983) Nucl. Acids Res. 11:2237–2255. The CAP protein (catabolite activator protein) is known to bind at certain promoter sites and to stimulate expression from that promoter: DeCrombrugghe, B., Busby, S. and Buc, H. (1984) Science 224: 831–838. Nucleotide sequences interspersed between these sites may not contribute to expression activity. Thus certain bases can be changed without affecting this activity. Also, the amino acid sequence of a protein is determined by the nucleotide sequence of the DNA. Because of the redundancy of the genetic code, i.e., more than one coding nucleotide triplet (codon) can be used for most of the amino acids used to make proteins, different nucleotide sequences can code for a particular amino acid. Thus, the genetic code can be depicted as follows:

Phenylalanine (Phe)	TTK	Histidine (His)	CAK
Leucine (Leu)	XTY	Glutamine (Gln)	CAJ

-continued

Isoleucine (Ile)	ATH	Asparagine (Asn)	AAK
Methionine (Met)	ATG	Lysine (Lys)	AAJ
Valine (Val)	GTL	Aspartic acid (Asp)	GAK
Serine (Ser)	QRS	Glutamic acid (Glu)	GAJ
Proline (Pro)	CCL	Cysteine (Cys)	TGK
Threonine (Thr)	ACL	Tryptophan (Try)	TGG
Alanine (Ala)	GCL	Arginine (Arg)	WGZ
Tyrosine (Tyr)	TAK	Glycine (Gly)	GGL
Termination signal	TAJ		
Termination signal	TGA		

Key: Each 3-letter deoxynucleotide triplet corresponds to a trinucleotide of mRNA, having a 5′-end on the left and a 3′-end on the right. All DNA sequences given herein are those of the strand whose sequence corresponds to the mRNA sequence, with thymine substituted for uracil. The letters stand for the purine or pyrimidine bases forming the deoxynucleotide sequence.

A=adenine
G=guanine
C=cytosine
T=thymine
X=T or C if Y is A or G
X=C if Y is C or T
Y=A, G, C or T if X is C
Y=A or G if X is T
W=C or A if Z is A or G
W=C if Z is C or T
Z=A, G, C or T if W is C
Z=A or G if W is A
QR=TC if S is A, G, C or T
J=A or G
K=T or C
L=A, T, C or G
M=A, C or T

The above shows that a novel amino acid sequence can be prepared by more than one nucleotide sequence. Accordingly, the subject invention includes such equivalent nucleotide sequences.

The work described herein was all done in conformity with physical and biological containment requirements specified in the NIH Guidelines.

What is claimed is:

1. A recombinant DNA high level expression vector comprising a first DNA segment consisting essentially of the following nucleotide sequence or equivalent nucleotide sequences containing bases whose translated region codes for the same amino acid sequence:

```
GATCTGACCT   ACGGTGTACT   GGCCGATATC
GAAGCGGAAG   ACCTGGCGCG   TGAAGCGTCG
TTTGCTCAGG   GATTACGCGC   GATGATTGGC
GGTATCTTAA   CCGCATCCTG   ATTCTCTCTC
TTTTTCGGCG   GGCTGGTGAT   AACTGTGCCC
GCGTTTCATA   TCGTAATTTC   TCTGTGCAAA
AATTATCCTT   CCCGGCTTCG   GAGAATTCCC
CCCAAAATAT   TCACTGTAGC   CATATGTCAT
GAGAGTTTAT   CGTTCCCAAT   ACGCTCGAAC
GAACGTTCGG   TTGCTTATTT   TATGGCTTCT
GTCAACGCTG   TTTTAAAGAT   TAATGCGATC
TATATCACGC   TGTGGGTATT   GCAGTTTTTG
GTTTTTTGAT   CGCGGTGTCA   GTTCTTTTTA
TTTCCATTTC   TCTTCCATGG   GTTTCTCACA
GATAACTGTG   TGCAACACAG   AATTGGTTAA
CTAATCAGAT   TAAAGGTTGA   CCAGTATTAT
TATCTTAATG   AGGAGTCCCT
  1                                       10
T ATG TTA CGT CCT GTA GAA ACC CCA ACC CGT
                                          20
GAA ATC AAA AAA CTC GAC GGC CTG TGG GCA TTC
```

-continued

30
AGT CTG GAT CGC GAA AAC TGT GGA ATT GAT
40
CAG CGT TGG TGG GAA AGC GCG TTA CAA GAA
50
AGC CGG GCA ATT GCT GTG CCA GGC AGT TTT
60
AAC GAT CAG TTC GCC GAT GCA GAT ATT CGT
70
AAT TAT GCG GGC AAC GTC TGG TAT CAG CGC
80
GAA GTC TTT ATA CCG AAA GGT TGG GCA GGC
90
CAG CGT ATC GTG CTG CGT TTC GAT GCG GTC
100
ACT CAT TAC GGC AAA GTG TGG GTC AAT AAT
110
CAG GAA GTG ATG GAG CAT CAG GGC GGC TAT
120
ACG CCA TTT GAA GCC GAT GTC ACG CCG TAT
130
GTT ATT GCC GGG AAA AGT GTA CGT ATC ACC
140
GTT TGT GTG AAC AAC GAA CTG AAC TGG CAG
150
ACT ATC CCG CCG GGA ATG GTG ATT ACC GAC
160
GAA AAC GGC AAG AAA AAG CAG TCT TAC TTC
CAT GAT TTC TTT AAC TAT GCC GGG ATC

and wherein said first segment is operably linked to a DNA sequence heterologous to said first segment.

2. The DNA transfer vector of claim 1 transferred to and replicated in a gram negative prokaryotic microorganism.

3. The DNA transfer vector of claim 2 wherein said gram negative prokaryotic microorganism is an *E. coli* K-12 derivative strain.

4. The amino acid sequence of the translated portion of the nucleotide sequence of the transfer vector of claim 1:

Met Leu Arg Pro Val Glu Thr Pro Thr Arg Glu Ile Lys
Lys Leu Asp Gly Leu Trp Ala Phe Ser Leu Asp Arg Glu
Asn Cys Gly Ile Asp Gln Arg Trp Trp Glu Ser Ala Leu
Gln Glu Ser Arg Ala Ile Ala Val Pro Gly Ser Phe Asn
Asp Gln Phe Ala Asp Ala Asp Ile Arg Asn Tyr Ala Gly
Asn Val Trp Tyr Gln Arg Glu Val Phe Ile Pro Lys Gly
Trp Ala Gly Gln Arg Ile Val Leu Arg Phe Asp Ala Val
Thr His Tyr Gly Lys Val Trp Val Asn Asn Gln Glu Val
Met Glu His Gln Gly Gly Tyr Thr Pro Phe Glu Ala Asp
Val Thr Pro Tyr Val Ile Ala Gly Lys Ser Val Arg Ile
Thr Val Cys Val Asn Asn Glu Leu Asn Trp Gln Thr Ile
Pro Pro Gly Met Val Ile Thr Asp Glu Asn Gly Lys Lys
Lys Gln Ser Tyr Phe His Asp Phe Phe Asn Tyr Ala Gly
Ile.

5. A recombinant DNA transfer vector, according to claim 1, comprising DNA having non-coding regions of DNA.

6. A recombinant DNA transfer vector which comprises DNA coding for a hybrid protein comprising DNA coding for part or all of the amino acids of *E. coli* β-glucuronidase wherein the *E. coli* β-glucuronidase DNA is fused to a DNA sequence coding for another protein.

7. A recombinant DNA transfer vector, according to claim 6, wherein said protein other than *E. coli* β-glucuronidase is a protein foreign to *E. coli.*

8. The DNA transfer vector of claim 6 transferred to and replicated in a gram negative prokaryotic microorganism.

9. The DNA transfer vector of claim 6 transferred to and replicated in a gram negative prokaryotic microorganism which then produces a protein containing part or all of the amino acid sequence of β-glucuronidase.

10. Plasmid pBG101-41 comprising the entire genome of pBR322 and having the restriction endonuclease pattern as shown in FIG. 1.

11. Plasmid pBG1 comprising the entire genome of pBR322 and having the restriction endonuclease pattern as shown in FIG. 1, and DNA having the following nucleotide sequence or equivalent nucleotide sequences containing bases whose translated region codes for the same amino acid sequence:

GATCTGACCT	ACGGTGTACT	GGCCGATATC
GAAGCGGAAG	ACCTGGCGCG	TGAAGCGTCG
TTTGCTCAGG	GATTACGCGC	GATGATTGGC
GGTATCTTAA	CCGCATCCTG	ATTCTCTCTC
TTTTTCGGCG	GGCTGGTGAT	AACTGTGCCC
GCGTTTCATA	TCGTAATTTC	TCTGTGCAAA
AATTATCCTT	CCCGGCTTCG	GAGAATTCCC
CCCAAAATAT	TCACTGTAGC	CATATGTCAT
GAGAGTTTAT	CGTTCCCAAT	ACGCTCGAAC
GAACGTTCGG	TTGCTTATTT	TATGGCTTCT
GTCAACGCTG	TTTTAAAGAT	TAATGCGATC
TATATCACGC	TGTGGGTATT	GCAGTTTTTG
GTTTTTTGAT	CGCGGTGTCA	GTTCTTTTTA
TTTCCATTTC	TCTTCCATGG	GTTTCTCACA
GATAACTGTG	TGCAACACAG	AATTGGTTAA
CTAATCAGAT	TAAAGGTTGA	CCAGTATTAT
TATCTTAATG	AGGAGTCCCT	

1 10
T ATG TTA CGT CCT GTA GAA ACC CCA ACC CGT
20
GAA ATC AAA AAA CTC GAC GGC CTG TGG GCA TTC
30
AGT CTG GAT CGC GAA AAC TGT GGA ATT GAT
40
CAG CGT TGG TGG GAA AGC GCG TTA CAA GAA
50
AGC CGG GCA ATT GCT GTG CCA GGC AGT TTT
60
AAC GAT CAG TTC GCC GAT GCA GAT ATT CGT
70
AAT TAT GCG GGC AAC GTC TGG TAT CAG CGC
80
GAA GTC TTT ATA CCG AAA GGT TGG GCA GGC
90
CAG CGT ATC GTG CTG CGT TTC GAT GCG GTC
100
ACT CAT TAC GGC AAA GTG TGG GTC AAT AAT
110
CAG GAA GTG ATG GAG CAT CAG GGC GGC TAT
120
ACG CCA TTT GAA GCC GAT GTC ACG CCG TAT
130
GTT ATT GCC GGG AAA AGT GTA CGT ATC ACC
140
GTT TGT GTG AAC AAC GAA CTG AAC TGG CAG
150
ACT ATC CCG CCG GGA ATG GTG ATT ACC GAC
160
GAA AAC GGC AAG AAA AAG CAG TCT TAC TTC
CAT GAT TTC TTT AAC TAT GCC GGG ATC.

12. A microorganism transformed by the transfer vector of claim 1.

13. A microorganism transformed by the transfer vector of claim 10.

14. *E. coli* MS371(pBG1), a microorganism according to claim 13.

15. *E. coli* MS371(pBG101-41), a microorganism according to claim 13.

16. A process for preparing recombinant plasmid pBG1 which comprises

(a) cutting plasmid pBG101-41 with restriction endonucleases SalI and BamHI to obtain a 1237 bp fragment; and

(b) ligating said 1237 bp fragment into a SalI, BamHI cut pBR322 vector.

17. A process for preparing *β*-glucuronidase which comprises culturing a gram negative prokaryotic microbe hosting plasmid pBG101-41.

18. A process, according to claim **17**, wherein said prokaryotic microbe is an *E. coli* K-12 derivative strain.

19. A process, according to claim **18**, wherein said *E. coli* is *E. coli* MS371(pBG101-41).

20. DNA having the following nucleotide sequence or equivalent nucleotide sequences coding for the same amino acid sequence:

```
GATCTGACCT   ACGGTGTACT   GGCCGATATC
GAAGCGGAAG   ACCTGGCGCG   TGAAGCGTCG
TTTGCTCAGG   GATTACGCGC   GATGATTGGC
GGTATCTTAA   CCGCATCCTG   ATTCTCTCTC
TTTTTCGGCG   GGCTGGTGAT   AACTGTGCCC
GCGTTTCATA   TCGTAATTTC   TCTGTGCAAA
AATTATCCTT   CCCGGCTTCG   GAGAATTCCC
CCCAAAATAT   TCACTGTAGC   CATATGTCAT
GAGAGTTTAT   CGTTCCCAAT   ACGCTCGAAC
GAACGTTCGG   TTGCTTATTT   TATGGCTTCT
GTCAACGCTG   TTTTAAAGAT   TAATGCGATC
TATATCACGC   TGTGGGTATT   GCAGTTTTTG
GTTTTTTGAT   CGCGGTGTCA   GTTCTTTTTA
TTTCCATTTC   TCTTCCATGG   GTTTCTCACA
GATAACTGTG   TGCAACACAG   AATTGGTTAA
CTAATCAGAT   TAAAGGTTGA   CCAGTATTAT
TATCTTAATG   AGGAGTCCCT
  1                                      10
T ATG TTA CGT CCT GTA GAA ACC CCA ACC CGT
                                         20
 GAA ATC AAA AAA CTC GAC GGC CTG TGG GCA TTC
                                     30
   AGT CTG GAT CGC GAA AAC TGT GGA ATT GAT
                                     40
   CAG CGT TGG TGG GAA AGC GCG TTA CAA GAA
                                     50
   AGC CGG GCA ATT GCT GTG CCA GGC AGT TTT
                                     60
   AAC GAT CAG TTC GCC GAT GCA GAT ATT CGT
```


-continued

```
                                  70
AAT TAT GCG GGC AAC GTC TGG TAT CAG CGC
                                  80
GAA GTC TTT ATA CCG AAA GGT TGG GCA GGC
                                  90
CAG CGT ATC GTG CTG CGT TTC GAT GCG GTC
                                  100
ACT CAT TAC GGC AAA GTG TGG GTC AAT AAT
                                  110
CAG GAA GTG ATG GAG CAT CAG GGC GGC TAT
                                  120
ACG CCA TTT GAA GCC GAT GTC ACG CCG TAT
                                  130
GTT ATT GCC GGG AAA AGT GTA CGT ATC ACC
                                  140
GTT TGT GTG AAC AAC GAA CTG AAC TGG CAG
                                  150
ACT ATC CCG CCG GGA ATG GTG ATT ACC GAC
                                  160
GAA AAC GGC AAG AAA AAG CAG TCT TAC TTC
        CAT GAT TTC TTT AAC TAT GCC GGG ATC.
```

21. DNA having the following nucleotide sequence or equivalent nucleotide sequences coding for the same amino acid sequence:

```
NdeI                                    TaqI
CATATGTCAT   GAGAGTTTAT   CGTTCCCAAT   ACGCTCGAAC

GAACGTTCGG   TTGCTTATTT   TATGGCTTCT   GTCAACGCTG

 AhaIII              Sau3A
TTTTAAAGAT   TAATGCGATC   TATATCACGC   TGTGGGTATT

                     Sau3A
GCAGTTTTTG   GTTTTTTGAT   CGCGGTGTCA   GTTCTTTTTA

                  NcoI
TTTCCATTTC   TCTTCCATGG   GTTTCTCACA   GATAACTGTG

                     HpaI
TCCAACACAG   AATTGGTTAA   CTAATCAGAT   TAAAGGTTGA

                          RBS              1
CCAGTATTAT   TATCTTAATG   AGGAGTCCCT   T   ATG   TTA   CGT

                                 10                 TaqI
CCT   GTA   GAA   ACC   CCA   ACC   CGT   GAA   ATC   AAA   AAA   CTC   GAC   GGC.
```

22. A recombinant DNA transfer vector comprising the DNA sequence of claim **21**.

23. A process for preparing useful hybrid proteins which comprises culturing a gram negative prokaryotic microbe hosting a recombinant DNA transfer vector comprising all or part of the *E. coli* *β*-glucuronidase structural gene DNA or promoter DNA.

24. A process, according to claim **23**, wherein said *E. coli* *β*-glucuronidase structural gene DNA or promoter DNA is obtained from an *E. coli* K-12 derivative strain.

25. A process, according to claim **24**, wherein said *E. coli* K-12 derivative is *E. coli* MS371.

26. A process, according to claim **24**, wherein said prokaryotic microbe is an *E. coli* K-12 derivative strain.

27. A process, according to claim **26**, wherein said *E. coli* K-12 derivative is *E. coli* MS371.

Chart A

```
Sau3A                                 TaqI
GATCTGACCT   ACGGTGTACT   GGCCGATATC   GAAGCGGAAG   ACCTGGCGCG

TGAAGCGTCG   TTTGCTCAGG   GATTACGCGC   GATGATTGGC   GGTATCTTAA   CCGCATCCTG

ATTCTCTCTC   TTTTTCGGCG   GGCTGGTGAT   AACTGTGCCC   GCGTTTCATA   TCGTAATTTC

                                       EcoRI
TCTGTGCAAA   AATTATCCTT   CCCGGCTTCG   GAGAATTCCC   CCCAAAATAT   TCACTGTAGC
```

17 18

-continued

Chart A

NdeI TaqI
CATATGTCAT GAGAGTTTAT CGTTCCCAAT ACGCTCGAAC GAACGTTCGG TTGCTTATTT

AhaIII Sau3A
TATGGCTTCT GTCAACGCTG TTTTAAAGAT TAATGCGATC TATATCACGC TGTGGGTATT

Sau3A NcoI
GCAGTTTTTG GTTTTTTGAT CGCGGTGTCA GTTCTTTTTA TTTCCATTTC TCTTCCATGG

HpaI
GTTTCTCACA GATAACTGTG TGCAACACAG AATTGGTTAA CTAATCAGAT TAAAGGTTGA

RBS 1
CCAGTATTAT TATCTTAATG AGGAGTCCCT T ATG TTA CGT CCT GTA GAA ACC CCA
MET LEU ARG PRO VAL GLU THR PRO

10 TaqI 20 Sau3A
ACC CGT GAA ATC AAA AAA CTC GAC GGC CTG TGG GCA TTC AGT CTG GAT CGC
THR ARG GLU ILE LYS LYS LEU ASP GLY LEU TRP ALA PHE SER LEU ASP ARG

BclI
30 Sau3A 40
GAA AAC TGT GGA ATT GAT CAG CGT TGG TGG GAA AGC GCG TTA CAA GAA AGC
GLU ASN CYS GLY ILE ASP GLN ARG TRP TRP GLU SER ALA LEU GLN GLU SER

50 Sau3A
CGG GCA ATT GCT GTG CCA GGC AGT TTT AAC GAT CAG TTC GCC GAT GCA GAT
ARG ALA ILE ALA VAL PRO GLY SER PHE ASN ASP GLN PHE ALA ASP ALA ASP

60 70
ATT CGT AAT TAT GCG GGC AAC GTC TGG TAT CAG CGC GAA GTC TTT ATA CCG
ILE ARG ASN TYR ALA GLY ASN VAL TRP TYR GLN ARG GLU VAL PHE ILE PRO

80 Taq I 90
AAA GGT TGG GCA GGC CAG CGT ATC GTG CTG CGT TTC GAT GCG GTC ACT CAT
LYS GLY TRP ALA GLY GLN ARG ILE VAL LEU ARG PHE ASP ALA VAL THR HIS

100 110
TAC GGC AAA GTG TGG GTC AAT AAT CAG GAA GTG ATG GAG CAT CAG GGC GGC
TYR GLY LYS VAL TRP VAL ASN ASN GLN GLU VAL MET GLU HIS GLN GLY GLY

120
TAT ACG CCA TTT GAA GCC GAT GTC ACG CCG TAT GTT ATT GCC GGG AAA AGT
TYR THR PRO PHE GLU ALA ASP VAL THR PRO TYR VAL ILE ALA GLY LYS SER

Rsa I 130 140
GTA CGT ATC ACC GTT TGT GTG AAC AAC GAA CTG AAC TGG CAG ACT ATC CCG
VAL ARG ILE THR VAL CYS VAL ASN ASN GLU LEU ASN TRP GLN THR ILE PRO

150 160
CCG GGA ATG GTG ATT ACC GAC GAA AAC GGC AAG AAA AAG CAG TCT TAC TTC
PRO GLY MET VAL ILE THR ASP GLU ASN GLY LYS LYS LYS GLN SER TYR PHE

BamHI
Sau3A
CAT GAT TTC TTT AAC TAT GCC GGG ATC
HIS ASP PHE PHE ASN TYR ALA GLY ILE

Supplementary Reading

Books and Articles

Crespi, R.S. 1982. *Patenting in the Biological Sciences: A Practical Guide for Research Scientists in Biotechnology and the Pharmaceutical and Agrochemical Industries.* John Wiley & Sons.

A general introduction to the various issues involved in patenting biological inventions. Includes a discussion of the patent systems used in various foreign countries.

Eisenberg, Rebecca S. 1987. Proprietary rights and the norms of science in biotechnology research. *Yale Law Journal* 97:177, December.

An examination of the influence that legal considerations can have on research and on the dissemination of research results. Although this article is not intended to provide guidance in protecting property rights and it is written primarily for legal audiences, the author does provide interesting insights into many topics relating to the law that are of interest to researchers.

Grubb, Philip W. 1986. *Patents in Chemistry and Biotechnology.* Clarendon Press.

A recently revised edition that has a much greater focus on biotechnology than the previous edition. The previous edition was entitled *Patents in Chemistry.* Covers a very broad range of topics ranging from the patent law history of Great Britain and the United States to discussions on licensing and infringement of patents.

Much of the book deals with patent procedures in Great Britain and other foreign countries.

Kanare, Howard M. 1985. *Writing the Laboratory Notebook*. American Chemical Society.

Provides detailed instructions on how to keep a laboratory notebook. The information is intended to be used by researchers not only to protect their legal property rights, but also to make their research more fruitful and efficient.

Lipscomb, Ernest B. III. 1985. *Lipscomb's Walker on Patents*. The Lawyers Co-operative Publishing Co. and Bancroft-Whitney Co., Third Edition.

An expansive multivolume treatise covering virtually all the general issues of patent law. An excellent reference source for technical aspects of patent law questions.

Patent Profiles—Biotechnology: 1982 Update. 1983. Office of Technology Assessment and Forecast.

A study of United States patents in several areas of biotechnology. In addition to information regarding the national and organizational origin of biotechnology patentees, this publication also contains a reprint of the first page of many United States biotechnology patents.

Saliwanchik, Roman. 1982. *Legal Protection for Microbiological and Genetic Engineering Inventions*. Addison-Wesley.

Written for patent attorneys. Focuses on many of the topics that are of interest to lawyers who prosecute patent applications for biotechnological inventions. Detailed analysis of specific legal arguments and court decisions is provided.

Source Book of Biotechnology Patents: 1981–1985. 1986. McGraw-Hill.

A compilation of issued United States patents, as well as published European and Japanese patent applications. This compilation provides the patent or application number, the names and countries of the inventors, the assignee, the filing or issue date, and a brief description of the subject matter of the patent or application. The information is indexed according to key word, company name, and patent number.

Periodicals and Patent News Services

By definition, the patenting of inventions involves rapidly changing areas of science and the law. To keep up with the many important events in this and related fields, it is helpful—if not essential—to peruse relevant periodicals and news services on a regular basis. Often, scientific periodicals refer to major patent law developments, but for more timely and extensive information, the following are useful:

BNA's Patent, Trademark & Copyright Journal.

A weekly newsletter covering all aspects of intellectual property. Although this newsletter does not focus specifically on biotechnology, much of the current news does relate to it.

Bioinvention.

A monthly journal published by OMEC International, Inc. Contains the titles and representative claims of recently issued United States patents for biotechnological inventions. The patents are conveniently organized according to subject matter. Also contains news briefs and licensing information, as well as assignee and inventor indexes.

Biotechnology Patentwatch.

Published every other week by McGraw-Hill, this newsletter gives the inventor, patent number, and a brief description of recently issued United States patents and recently published European and Japanese patent application disclosures.

PRI-Opinions.

A weekly compilation of the published opinions of the Court of Appeals for the Federal Circuit. This publication is primarily of interest to patent attorneys and others with a need for in-depth knowledge of current appellate court decisions relating to patent law.

Index